SOUTHAMPTON TECHNICAL COLLEGE
DEPT.
MATHS AND SCIENCE

WITHDRAWN FOR REASON BELOW

ACC NO : 021926-6
TITLE : CONSTRUCTION MATHEMATICS
VOL 2
DDC(1) : 510

Construction Mathematics

Volume 2

Construction Mathematics

Volume 2

BRIAN W. BOUGHTON
C.Eng., M.I. Struct.E.
Educational Advisory Officer for Construction
Business and Technician Education Council

and

PETER J. BALLARD
B.Ed., M.I.S.M.E.
Former Senior Administrative Officer for
Construction and Science,
Business and Technician Education Council

COLLINS
8 Grafton Street, London W1

Collins Professional and Technical Books
William Collins Sons & Co. Ltd
8 Grafton Street, London W1X 3LA

First published in Great Britain by
Collins Professional and Technical Books 1985

Distributed in the United States of America
by Sheridan House, Inc.

British Library Cataloguing in Publication Data
Boughton, Brian W.
Construction mathematics.
Vol. 2
1. Building–Mathematics
I. Title II. Ballard, Peter J.
510'.24624 TH153

ISBN 0 00 383035 7

Typeset by Cambrian Typesetters, Frimley, Surrey
Printed and bound in Great Britain by
Mackays of Chatham, Kent

Contents

Preface

This book is intended, primarily, to cover the content of the BTEC units:

Analytical Mathematics II
Mathematics III, and
Statistics III

together with elements of:

Use of Computers II, and
Site Surveying and Levelling II.

It does, however, overlap with some of the science-based units by virtue of their dependence on an understanding of mathematical techniques. The use of electronic calculators is developed, from that shown in Volume 1, to include programmable calculators, which act as a useful introduction to computing. As with Volume 1, the mathematical theories are dealt with prior to illustrating their application, where possible, to the solution of constructional problems.

Each chapter contains exercises to develop self-assessment, and some include assignment work to show the value of mathematics as a construction 'tool'.

This book is not intended to produce mathematicians or scientists but is aimed more directly at technicians within the construction sector who need to develop a level of confidence in the use of the theories, principles and techniques as a means of solving the types of problem likely to be encountered in the working environment.

1 Geometry and Trigonometry

In Volume 1 we considered the properties of the complete circle, but in construction, particularly surveying for civil engineering, we are more often concerned with only a portion of a circle. Let us look, therefore, at the properties of a circular curve that connects two straight lines, such as might be encountered in the setting out of a road or railway layout.

GEOMETRY OF THE CIRCULAR CURVE

Figure 1.1 shows a circular curve *ab* connecting two straight lines which are produced to intersect at a point *d*. These are tangents to the circle of centre *o* and radius *R* as shown, and the radii are produced to form right angles $o\hat{a}d$ and $o\hat{b}d$. Line *ad* is also extended to d_1 to produce angle $b\hat{d}d_1$ which is known as the *deflection angle* of the curve.

If we now bisect the angle subtended by the curve $a\hat{o}b$, we can see that it will also intersect point *d*, the intersection of the tangents, so that triangles *oad* and *obd* are congruent. It follows that since the angles $a\hat{o}b$ and $a\hat{d}b$ summate to 180° (total angles of a quadrilateral equal 360°), then the subtended angle to the curve must equal its deflection angle. Thus $a\hat{o}b = b\hat{d}d_1$.

We now connect points *a* and *b* by a straight line to form the chord to the curve which cuts the bisecting line *od* at *f*, producing four right-angled triangles. If we now bisect the angles $d\hat{a}f$ and $d\hat{b}f$, we will find that these bisecting lines cross line *od* where it cuts the curve at point *c*. This is known as the *crown point* of the curve. The resultant triangles *afc* and *bfc* are congruent.

If we examine the resultant sets of triangles reflected about line *ofcd*, as shown in fig. 1.1, we can see that angles $a\hat{o}d$ and $a\hat{d}o$ summate to 90° (right angle at tangent to the circle); then angle

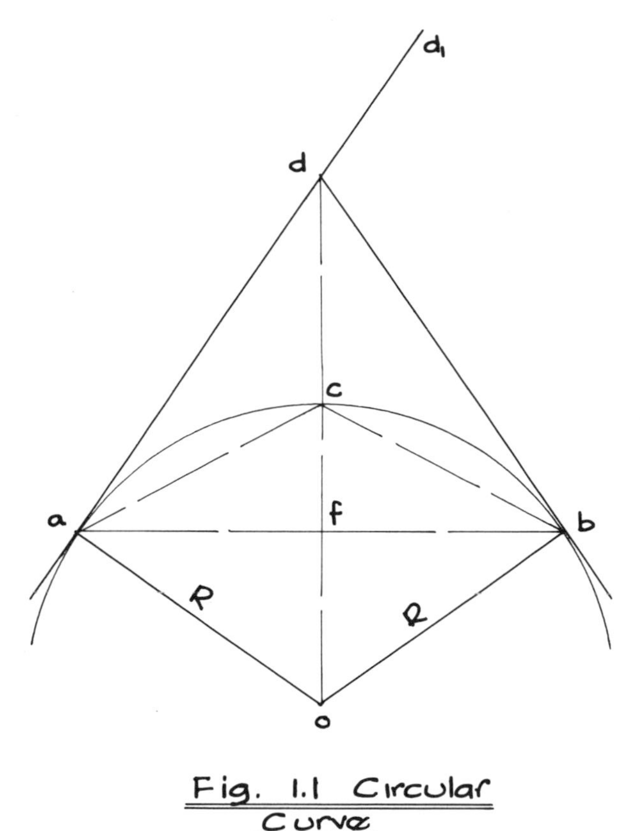

Fig. 1.1 Circular Curve

$d\hat{a}f$ must be equal to angle $a\hat{o}d$ (angle $a\hat{f}d$ is a right angle). Since we also bisected angles $d\hat{a}f$ and $a\hat{o}b$ in constructing the figure, then $d\hat{a}c = c\hat{a}f = \frac{1}{2}\, a\hat{o}d$, and $a\hat{o}d = d\hat{a}f = \frac{1}{2}\, b\hat{d}d_1$.

These co-related values, together with some annotations to distances, are given in fig. 1.2, which is the same shape as fig. 1.1 but with properties as used in surveying.

If we now consider another arc, knowing the values established above but looking at another property, we can see that if the chord to the arc is extended, its angle has the same relationship as shown in fig. 1.2 (where $d\hat{a}b = \frac{1}{2}\, a\hat{o}b$). This is shown in fig. 1.3, which shows only one tangent, the chord and its radii so that this angle δ is half that subtended by the arc which is, therefore, 2δ.

Having learnt in Volume 1 that the length of an arc is radius × angle subtended (in radians), we know that if the length of the arc is l then $2\delta \times R = l$, so that $\delta = l/2R$ radians, or, given that 1 radian $= 360/2\pi$ degrees, $\delta = 28.6364\ l/R$ degrees. This holds good for any

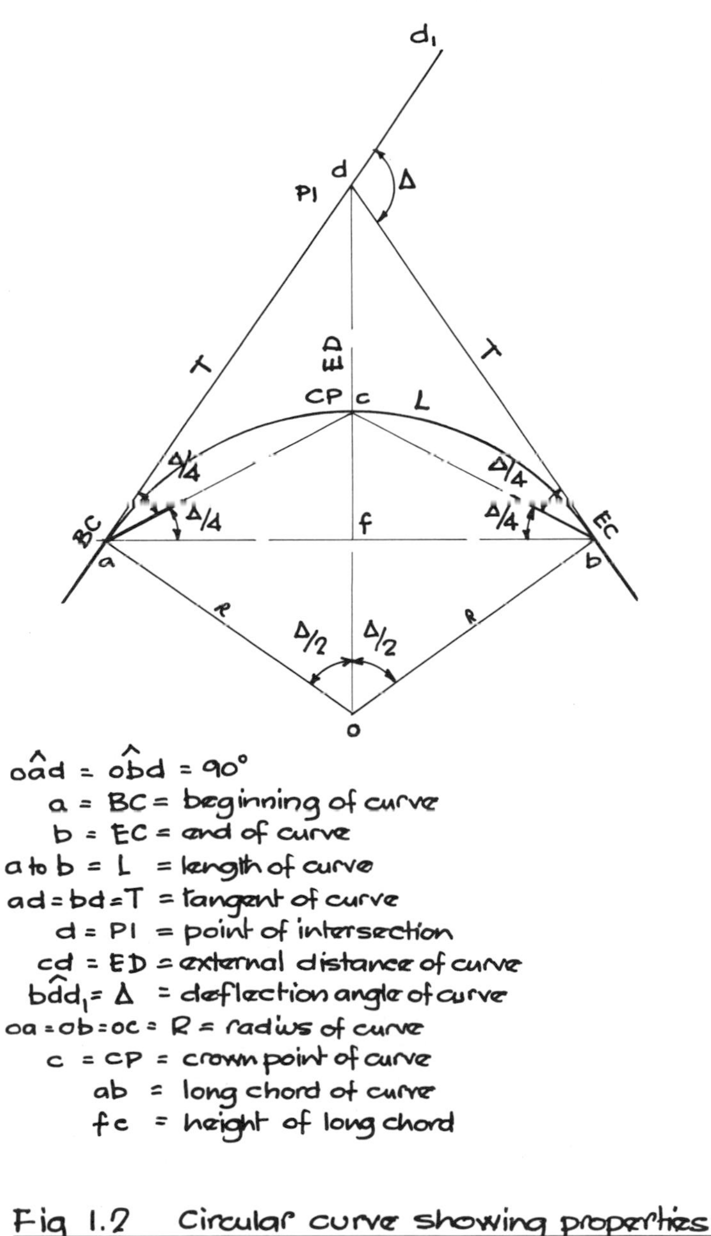

Fig 1.2 Circular curve showing properties used in surveying

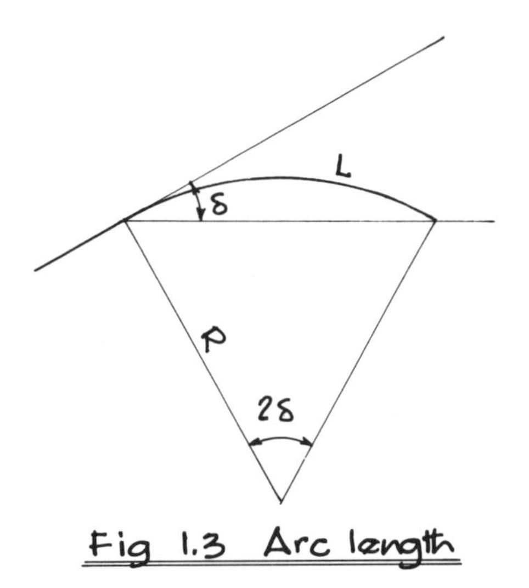

Fig 1.3 Arc length

arc/chord deflection angle, and is used extensively in surveying problems.

RECIPROCAL TRIGONOMETRIC RATIOS

Although we already know how to obtain the trigonometric values of sin, cos and tan, it is often useful to be able to apply their reciprocals and, rather than be forced to divide by quite awkward values, mathematical tables are available to give these values. The reciprocals are:

$$\frac{1}{\sin\theta} = \text{cosec}\,\theta \text{ (cosecant)}$$

$$\frac{1}{\cos\theta} = \sec\theta \text{ (secant)}$$

$$\frac{1}{\tan\theta} = \cot\theta \text{ (cotangent)}$$

Returning to fig. 1.2 we may now establish some trigonometric values of particular use in surveying, thus:

(a) $T/R = ad/ao = \tan \Delta/2$

therefore

$$T = R \tan \Delta/2 \qquad \text{(i)}$$

(b) $(oc + cd)/ao = \sec \Delta/2$, but since oc and ao are both radii to the curve and cd is the external distance ED, then $(\text{ED} + R)/R = \sec \Delta/2$, so that $\text{ED} + R = R \sec \Delta/2$

$$\text{ED} = R \sec \Delta/2 - R$$
$$= R (\sec \Delta/2 - 1) \qquad \text{(ii)}$$

(c) $of/oa = \cos \Delta/2$, but $of = oc - cf$, therefore $(oc - cf)/oa = \cos \Delta/2$, and since oc and oa are radii to the curve, and cf is the height of the chord

$$(R - \text{Ht of chord})/R = \cos \Delta/2$$
$$R - \text{Ht of chord} = R \cos \Delta/2$$
$$R - R \cos \Delta/2 = \text{Ht of chord}$$
$$R (1 - \cos \Delta/2) = \text{Ht of chord}$$

or Ht of chord $= R (1 - \cos \Delta/2)$ (iii)

(d) $af/ao = \sin \Delta/2$, and since ao is the radius to the curve, $af/R = \sin \Delta/2$. Thus ab, the chord length, being twice af, $ab/R = 2 \sin \Delta/2$

$ab = R \times 2 \sin \Delta/2$
or $2R \sin \Delta/2$

thus chord length $= 2R \sin \Delta/2$ (iv)

Note: for expressions (ii) and (iii) it is permissible to use an alternative expression to remove the numerical constant, i.e. $(\sec \alpha - 1)$ is the *exsecant* of α (exsec) and $(1 - \cos \alpha)$ is the *versed sine* of α (versin), so the expressions can be rephrased as:

$$\text{ED} = R \text{ exsec } \Delta/2$$
$$\text{and Ht of chord} = R \text{ versin } \Delta/2$$

These trigonometric values are available in some mathematical and surveying tables. If using an electronic calculator, however, such values are *not* available, and only the three basic values can be found directly, which would involve use of the $1/x$ reciprocal key. This is reasonably straightforward for expression (ii). For example, for a radius of 25 m and a deflection angle to the curve of 100°, the external distance, ED, is 25 (sec 50° - 1), thus enter 50 → cos → $1/x$ → - 1 → = → x 25 = 13.893 m. However, it is slightly more tortuous for expression (iii). For example, for the same radius and deflection angle, the height of the chord is 25 (1 - cos 50°), thus enter 50 → cos → +/- → + 1 → = → x 25 = 8.930 m.

TRIGONOMETRIC IDENTITIES

When we talked of the exsecant and versed sine, the expressions that they represented were, in fact, trigonometric identities. There are many of these but some of the more useful identities can be determined quite readily by use of a right-angled triangle as shown in fig. 1.4.

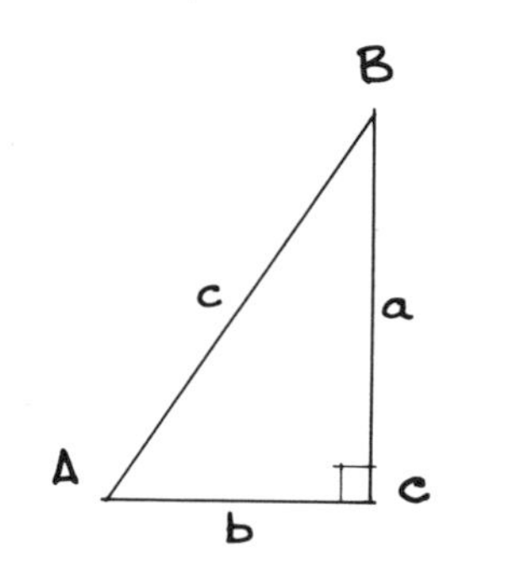

Fig 1.4 Trigonometric identities

We already know that, for angle A, $\sin A = a/c$, $\cos A = b/c$, and $\tan A = a/b$. So, if we divide $\sin A$ by $\cos A$, we will produce $a/c \div b/c$ or $a/c \times c/b$, which equals $a/b = \tan A$. Therefore:

$$\tan = \frac{\sin}{\cos} \qquad \text{(i)}$$

Having talked earlier of reciprocals, then

$$\text{cotan} = \frac{\cos}{\sin} \qquad \text{(ii)}$$

Returning to the basic ratios, if we square the sine and cosine values and add them together, we obtain

$$\sin^2 A = \frac{a^2}{c^2}, \quad \cos^2 A = \frac{b^2}{c^2}$$

so that $\sin^2 A + \cos^2 A = \dfrac{a^2 + b^2}{c^2}$

but we know, by using the Pythagoras theorem, that in a right-angled triangle (like that in fig. 1.4) $a^2 + b^2 = c^2$. Thus $\sin^2 A + \cos^2 A = c^2/c^2 = 1$. Therefore:

$$\sin^2 + \cos^2 = 1 \tag{iii}$$

which may be transposed to read either $\sin^2 = 1 - \cos^2$ or $\cos^2 = 1 - \sin^2$.

Now, if we square the tan value, we obtain

$$\tan^2 = \sin^2/\cos^2$$

but since $\sin^2 = 1 - \cos^2$

$$\tan^2 = (1 - \cos^2)/\cos^2, \qquad \text{or} \qquad (1/\cos^2) - 1$$
$$\tan^2 = \sec^2 - 1 \tag{iv}$$

Also, if we square the cotan value, we obtain

$$\text{cotan}^2 = \cos^2/\sin^2,$$

but since $\cos^2 = 1 - \sin^2$,

$$\text{cotan}^2 = (1 - \sin^2)/\sin^2 \qquad \text{or} \qquad (1/\sin^2) - 1$$

$$\text{cotan}^2 = \text{cosec}^2 - 1 \tag{v}$$

These expressions may also be transposed to read:

$$\sec^2 = 1 + \tan^2 \qquad \text{and} \qquad \text{cosec}^2 = 1 + \text{cotan}^2$$

MULTIPLE AND SUB-MULTIPLE ANGLES

Earlier in this chapter we talked about angles being twice as large, or half the size, of others. In order to determine ratios for such expressions we need to utilise the equations:

$$\sin(A \pm B) = \sin A \cos B \pm \cos A \sin B$$
and $$\cos(A \pm B) = \cos A \cos B \mp \sin A \sin B$$

First, let us determine the double value $2A$, thus:

$$\sin(A + A) = \sin A \cos A + \cos A \sin A$$
$$\sin 2A = 2(\sin A \cos A) \tag{i}$$

$$\cos(A + A) = \cos A \cos A - \sin A \sin A$$
$$\cos 2A = \cos^2 A - \sin^2 A \qquad \text{(ii)}$$

Because $\cos^2 + \sin^2 = 1$, the expression may be rewritten either as $\cos 2A = 2\cos^2 A - 1$ or $\cos 2A = 1 - 2\sin^2 A$. Also, since we know that tan = sin/cos, it follows that:

$$\tan(A + B) = \frac{\sin A \cos B + \cos A \sin B}{\cos A \cos B - \sin A \sin B}$$

This is rather cumbersome as an expression and can be simplified by dividing *everything* first by cos A and then by cos B. *Note:* this means *each* of the four components because of the + and - signs. (Normally, if we divide the top and bottom values of a fraction, or multiply them, we can see that it does not change the overall value. We are factorising! Refer back to Volume 1.)

$$\text{So divide by } \cos A = \frac{\tan A \cos B + \sin B}{\cos B - \tan A \sin B}$$

$$\text{Now divide by } \cos B = \frac{\tan A + \tan B}{1 - \tan A \tan B}$$

$$\text{thus} \quad \tan(A \pm B) = \frac{\tan A \pm \tan B}{1 \mp \tan A \tan B}$$

$$\text{so that} \quad \tan 2A = \frac{2\tan A}{1 - \tan^2 A} \qquad \text{(iii)}$$

We can follow the same process for half angles if we assume the value as $(\theta/2 + \theta/2)$ and obtain

$$\sin\theta = 2(\sin\theta/2 \cos\theta/2) \qquad \text{(iv)}$$

$$\cos\theta = \cos^2\theta/2 - \sin^2\theta/2 \qquad \text{(v)}$$

$$\tan\theta = \frac{2\tan\theta/2}{1 - \tan^2\theta/2} \qquad \text{(vi)}$$

In themselves these identities are not particularly useful, but they can be extremely valuable in solving trigonometric equations when they occur, for example, in sound-wave calculations.

CASE STUDY 1.1: APPLICATION OF GEOMETRY AND TRIGONOMETRY TO A ROAD OF CONSTANT INCLINE WITH CHANGE OF DIRECTION

A road climbs at a constant gradient of 8% and changes direction as shown in fig. 1.5. The lines in the diagram represent a plan view of the centre line of the road, and the tangent at which the curve will link the straight sections is joined by its chord. Given a carriage width between curbs of 10 m we want to determine the surface area within the run of road shown.

Since the gradient is constant, we can work on the 'plan' dimensions throughout until the final calculation to convert to 'true' length. We know that the deflection angle is 82° and the chord length is 42 m, so we can find the radius of the curve linking the two straights from:

$$\text{chord length} = 2R \sin \Delta/2$$

By transposing the equation to make R the subject, thus:

$$R = \frac{\text{chord length}}{2 \sin \Delta/2}$$

$$R = \frac{42\,000}{2 \sin 41°}$$

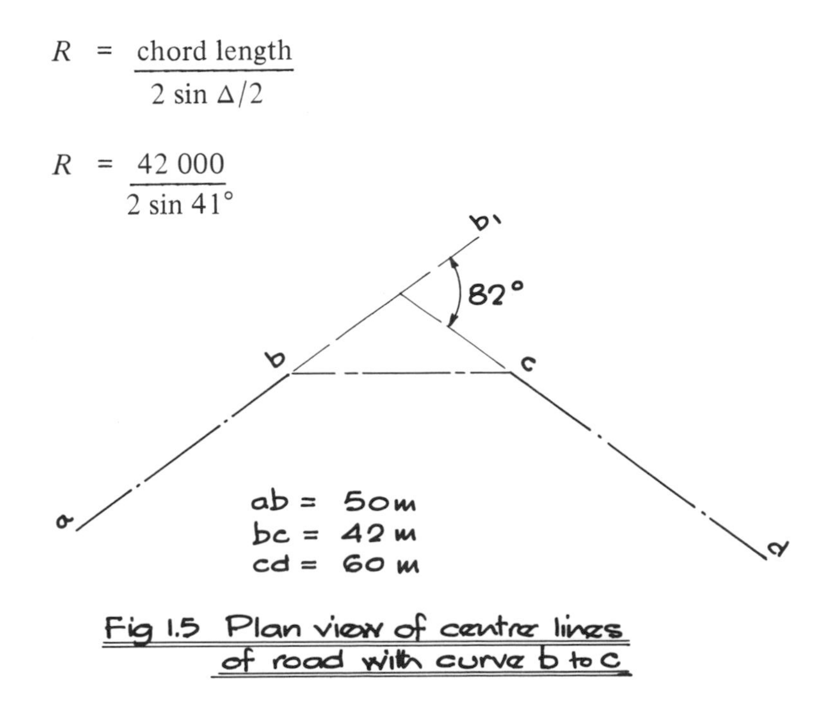

Fig 1.5 Plan view of centre lines of road with curve b to c

Using the calculator to determine R:

Enter 41 → sin → x 2 → = → $1/x$ → x 42 → = 32.009 m

To find the length of the arc we use the expression: subtended angle (in radians) x radius. We know, from earlier in this chapter, that the subtended angle is equal to the deflection angle, so arc length = 82° (converted to radians) x 32.009

Using the calculator to determine arc length:

Enter π → ÷ → 180 → x 82 → x 32.009 → = 45.810 m

Thus plan distance travelled at centre line = 50 + 45.81 + 60 = 155.81 m, so plan area of road surface = 155.81 x 10 = 1558.1 m^2 (theorem of Pappus).

Since the gradient of the road is an alternative way of expressing the tangent of its inclined angle, if $\tan \alpha = 0.0800$, $\alpha = 4.5739°$. We also know, from Volume 1, that the true area is equal to plan area/cos α. In practice, to avoid writing down the intermediate stage of calculating the angle we would have carried out both the tangent calculation and that involving the cosine, in sequence, thus:

Enter 0.08 → arc tan → cos → $1/x$ → x 1558.1 → = 1563.078 m^2

Because of the small amount in excess, we would use the figure 1563 m^2. You can, however, see the benefit of using the calculator, since we would otherwise have had to determine the cosine, or if preferred the secant, of 4.5739°.

CASE STUDY 1.2: TRIGONOMETRY RELATED TO TACHEOMETRY

In tacheometry a theodolite with stadia lines is used to determine distances by angular means. These lines are so arranged as to form the base of an isosceles triangle whose vertical height is exactly 100 times its base. Since they are stadia lines that form the base, its dimension is called s, which gives a length $d = 100s$, as shown in fig. 1.6. This is fine so long as the sighting is taken on the horizontal plane, but this is very rarely the case because the ground often slopes. Let us examine how we can use some of the trigonometry we have learnt to solve a problem using tacheometry on sloping ground.

Figure 1.7 shows a typical situation where the sighting is taken

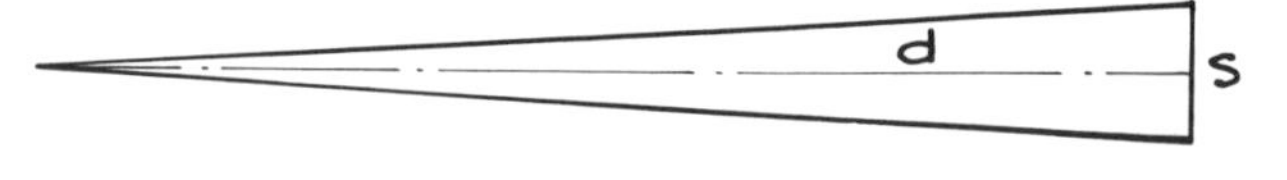

Fig 1.6 Stadia lines

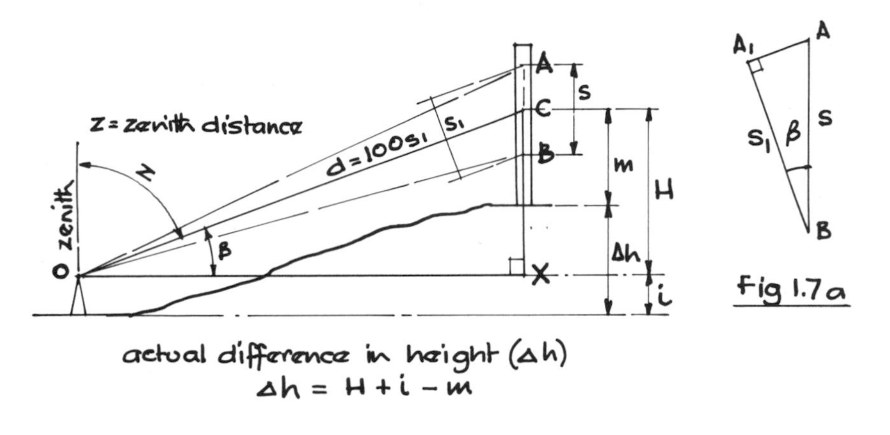

Fig 1.7 Sighting from a lower level

from a lower level than that of the staff. We need to find both the horizontal distance and the difference in level, or height, of the reading. Because the base of the isosceles triangle is at right angles to the sighting line, the distance d is not 100 x AB (or s) but is, instead, 100 x s_1, which is a reduced distance, as seen in the 'exploded detail', fig. 1.7a. Since the triangle has the same angles as triangle OCX then s_1 must be $s \times \cos\beta$ so that $d = 100s \cos\beta$. Similarly, the horizontal distance D must be $d \times \cos\beta$ so that $D = 100s \cos^2\beta$. This is a constant from which calculations can be made since the horizontal distance is related precisely to the angle of inclination.

Now, to determine the difference in level, we can use the same major triangle where $H = d \times \sin\beta$, so that $H = 100s \cos\beta \sin\beta$. Having looked at sub-multiple angles earlier in the chapter, we can see that this relates to $\sin\theta$, where, if we replace $\theta/2$ by β, we have: $\sin(\beta + \beta) = 2(\sin\beta \cos\beta)$, so that $\frac{1}{2}\sin 2\beta = \sin\beta \cos\beta$, giving us an alternative expression: $H = 100s\ \frac{1}{2}\sin 2\beta$, again this is a constant since the height is obviously related directly to the angle of inclination. If we apply these expressions to an actual case, it will show how easy the calculation becomes.

Taking the set-up of the theodolite to be as in fig. 1.7, with an angle β of 18.625°, the reading taken on the staff was: lower stadia line 1.166 m, upper line 1.384 m, giving a value for s of 1.384–1.166 = 0.218 m. This gives a value for $100s$ of 21.8 m. Thus horizontal distance $D = 21.8 \times \cos^2 18.625°$, and change in level $H = 21.8 \times \frac{1}{2}\sin 2 \times 18.625°$. Using a calculator for D:

insert 18.625 → cos → x^2 → x 21.8 → = 19.576 m
and for H:

insert 18.625 → x 2 → = → sin → ÷ 2 → = → x 21.8 = 6.598 m.

Note: in the second calculation it is important to realise that the *angle* is doubled (2β) but its *sine* value is halved (½ sin). The ½ and 2 do *not* cancel each other out since they are referring to *different* items.

TRIGONOMETRIC EQUATIONS

As mentioned earlier it is sometimes possible to solve what appear to be very complex trigonometric equations by making use of trigonometric identities to reduce the equations to expressions involving only one unknown.

Example 1.1

$12 \cos^2 x - \sin x - 11 = 0$. Determine x.

In order to eliminate the cos value we make use of the identity $\sin^2 x + \cos^2 x = 1$, therefore $\cos^2 x = 1 - \sin^2 x$, which, when substituted into the equation, produces: $12 - 12 \sin^2 x - \sin x - 11 = 0$, which, when collated to produce a positive value for $\sin^2 x$, gives $0 = 12 \sin^2 x + \sin x - 1$. This factorises to:

$$0 = (4 \sin x - 1)(3 \sin x + 1)$$

which gives values $\sin x = +0.2500$ or -0.3333.

Remembering that similar values occur for positive in the first two quadrants, and for negative in the last two quadrants, of the circle (Volume 1, figs 8.6 and 8.7), the angles are: for 0.2500, $x = 14.48°$ or $165.52°$, for -0.3333 $x = 199.47°$ or $340.53°$.

Example 1.2

$0 = 4 \sec \alpha \tan \alpha - 5 \operatorname{cosec} \alpha \cot \alpha$

We have to transpose the equation in order to establish a numerical value, i.e. $5 \operatorname{cosec} \alpha \cot \alpha = 4 \sec \alpha \tan \alpha$

thus $$\frac{5 \operatorname{cosec} \alpha \cot \alpha}{4 \sec \alpha \tan \alpha} = 1, \text{ or } \frac{4 \sec \alpha \tan \alpha}{5 \operatorname{cosec} \alpha \cot \alpha} = 1$$

Either equation will give the same result, but in order to reduce the number of unknowns we must eliminate by substitution. Taking the first equation

$$\operatorname{cosec} \alpha = \frac{1}{\sin \alpha} \quad \text{and} \quad \sec \alpha = \frac{1}{\cos \alpha}$$

so $$\frac{\operatorname{cosec} \alpha}{\sec \alpha} = \frac{\cos \alpha}{\sin \alpha} = \cot \alpha$$

also $$\frac{1}{\tan \alpha} = \cot \alpha$$

so the expression now reads $5/4 \cot \alpha . \cot \alpha . \cot \alpha = 1$
or $\cot^3 \alpha = 0.8000$
$\cot \alpha = 0.92832$
Alternatively, taking the second equation,

$$\frac{\sec \alpha}{\operatorname{cosec} \alpha} = \frac{\sin \alpha}{\cos \alpha} = \tan \alpha$$

and $$\frac{1}{\cot \alpha} = \tan \alpha,$$

so the expression reads $4/5 \tan\alpha . \tan\alpha . \tan\alpha = 1$, or $\tan^3 \alpha = 1.2500$
$\tan \alpha = 1.07722$
which is the reciprocal of 0.92832.
Thus $\alpha = 47.13°$ or $227.13°$.

Exercises
1.1 Two radii to a circle are connected by a chord of length 2.0 m, as shown in fig. 1.8. A tangent, produced through A, meets the radius, extended through B, at point C. The measured dimensions are $AC = 2.42$ m, $BC = 0.82$ m. From this information determine: (a) the radius of the circle, (b) the length of the subtended arc. (*Note*: the use of the cosine rule would simplify the early stages of the calculation.)

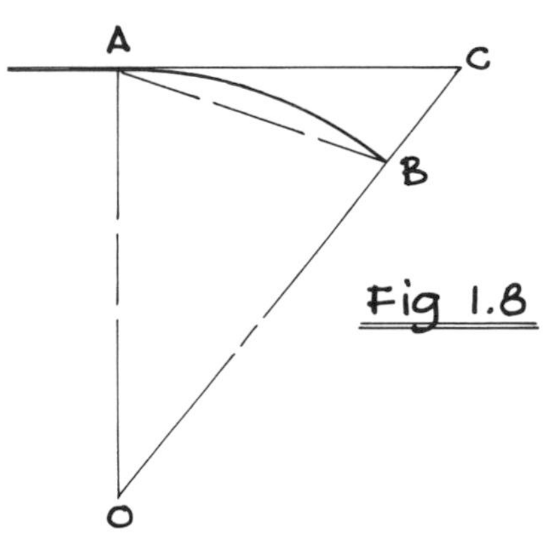

Fig 1.8

1.2 A concrete shell roof to an exhibition hall follows the form of a circular arc, as shown in fig. 1.9. The supports are raked struts at a tangent to the curve as indicated. Given an eaves height of 2.8 m, a chord length of 12 m and a central roof height of 4.5 m from floor level, determine (a) the radius of the curve, (b) the sectional angle of the support struts, (c) the distance from the eaves at which the struts enter the floor slab.

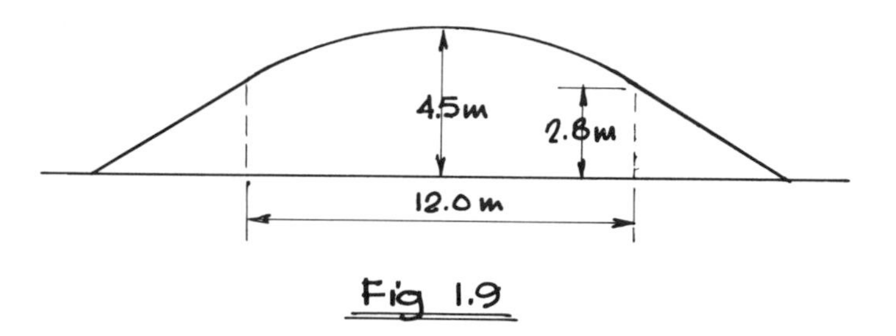

Fig 1.9

1.3 A theodolite with stadia lines is used to take sightings in a depression, as shown in fig. 1.10. The staff readings from point 1, for an angle 16.332°, are 1.615 m and 1.430 m, and from point 2, for an angle of 14.853°, are 1.520 m and 1.352 m. Determine: (a) the overall plan length from point 1 to the staff position for point 2 (*A* to *C*), and (b) the overall difference in ground level, between *A* and *C*, given a height of axis for the theodolite of 1.5 m.

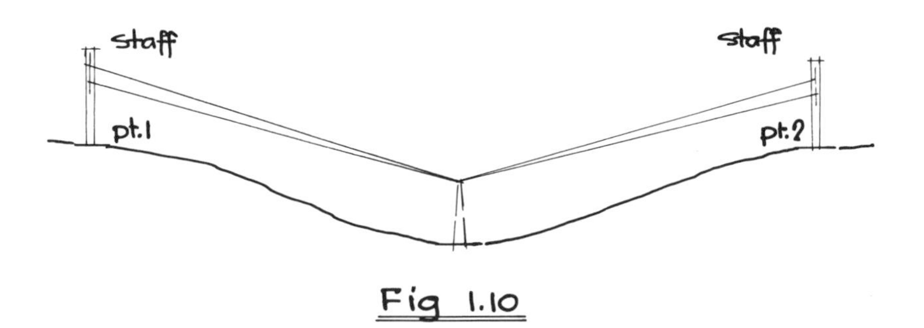

Fig 1.10

1.4 An alternative method for setting out a curve, using small equal-length chords, is by 'offsets' as shown in fig. 1.11. Knowing the radius of the curve is 12.00 m and its subtended angle is 60°, determine: (a) the length of the major chord, (b) the lengths of the five equal minor chords, and (c) their offsets in terms of x and y and x_1 to x_5 and y_1 to y_5.

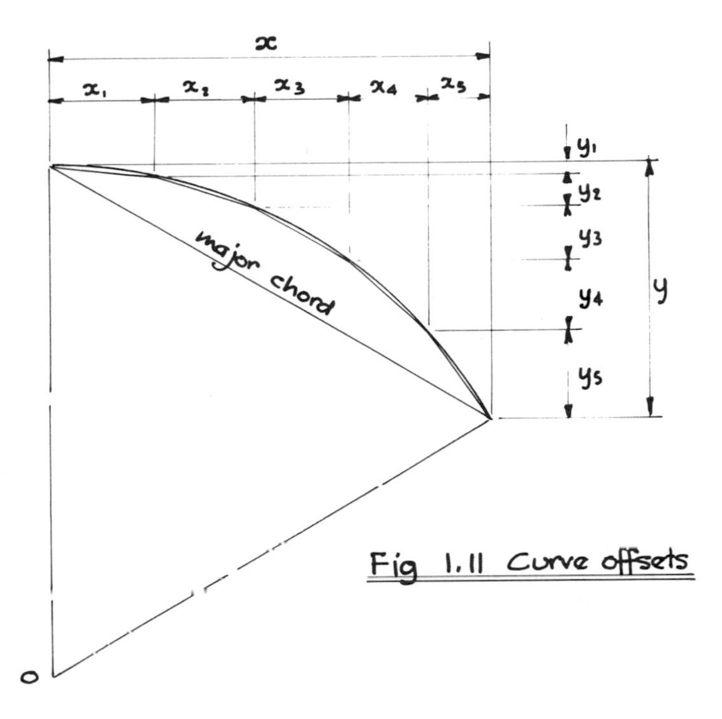

Fig 1.11 Curve offsets

1.5 (a) Given the trigonometric equation $\sec^2 \alpha = 2 \tan^2 \alpha$, solve for values of α between 0° and 360°.
(b) Given the Trigonometric equation $3 \sin^2 \theta = \cos \theta - 1$, solve for values of θ between 0° and 180°.

Assignment 1.1
Vertical curves used in road layouts, unlike horizontal curves, are not of circular format but are parabolic, with the 'external distance' being equal to the height of the 'main chord' (refer back to fig. 1.2 for comparison). This is to provide a more gentle transition from an uphill to a downhill slope, e.g. to avoid the 'hump-back bridge' effect. Determine the geometry involved in setting out such curves and explain how this may be translated in practice. (*Note:* reference to a book on surveying or highway design will help you to develop your answer.)

2 Mensuration

Although the elements of a building will normally possess a geometrical shape, or a combination of such shapes, the environment in which we build is most unlikely to follow any such pattern. Thus the areas and volumes that we considered in Volume 1 cannot be employed directly when considering, for example, the volume of earth to be removed from an irregular level site on which we wish to construct a basement.

IRREGULAR SHAPES

For plan areas of irregular shapes, one of the easiest ways to determine the overall area is to draw it to scale on 'squared' or graph paper. We can then summate the number of whole squares and make approximations with regard to the part squares in producing a reasonably accurate result. Remember, in construction it is not always necessary to determine *precise* figures, since we are dealing with the likelihood of tolerances in production. This is particularly true in earth-moving on a large scale, where the earth is prone to 'bulking' when disturbed, since we 'round up' to the next lorry load for carting away.

Example 2.1
Determine the plan area of an artificial lake to be constructed for a new golf course, as shown in fig. 2.1. By scaling the drawing we can establish that the approximate maximum dimensions of the lake are 59 m x 37 m so that its shape can be contained within a 6 x 4 grid with each square divided into tenths. This can be achieved by using tracing paper to transfer the shape from the drawing to a graph sheet (see fig. 2.2).

If we consider the 5 x 5 squares, there are 39 whole (or very nearly whole) squares. If we now look at the remainder, $a + b$ is

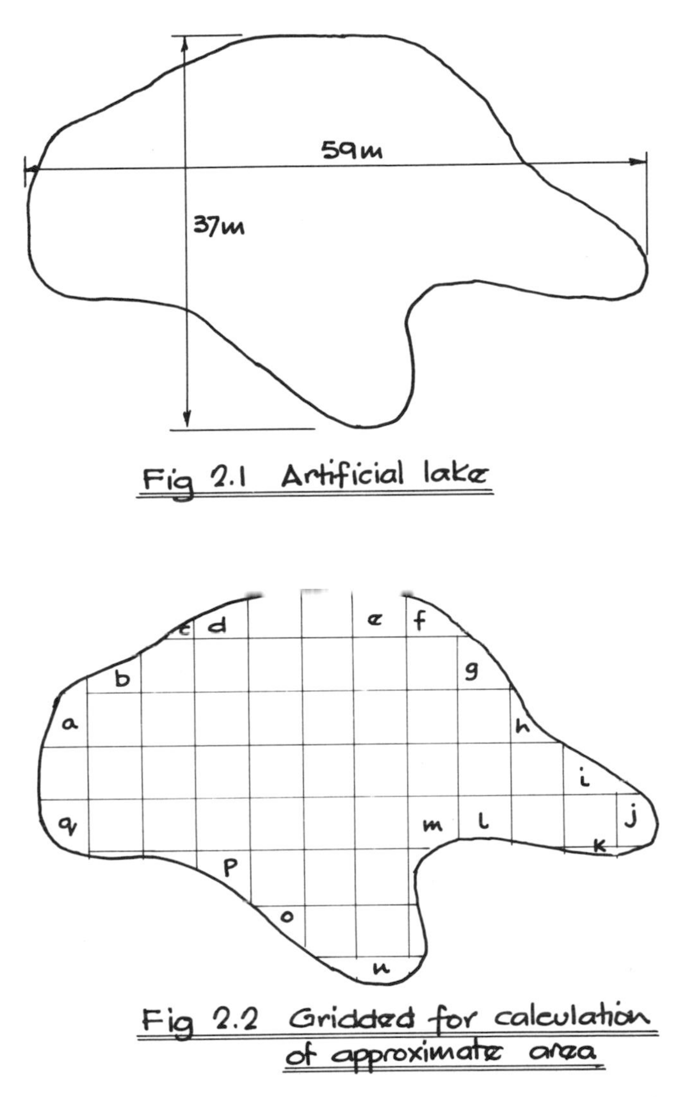

Fig 2.1 Artificial lake

Fig 2.2 Gridded for calculation of approximate area

approximately 1½ squares, $c + d$ is approximately 1 square, $e + f + g$ is approximately 2 squares, $h + i$ is approximately 1 square, $j + k$ is approximately 1 square, $l + m + n$ is approximately 3 squares, and $o + p + q$ is approximately 1⅔ squares. Thus the overall total is: 39 + 1.5 + 1 + 2 + 1 + 1 + 3 + 1.66 = 50.16 squares, or 50.16 x 25 = 1254 m². We would then round this up to 1260 m².

An alternative technique is to divide the shape into equal width strips and summate their middle lengths. This is known as the mid-ordinate rule.

MID-ORDINATE RULE

As explained above, when a shape is divided into equal-width strips, each area is approximately width x ordinate (see fig. 2.3) so that the total area equals width of strip x total length of ordinates.

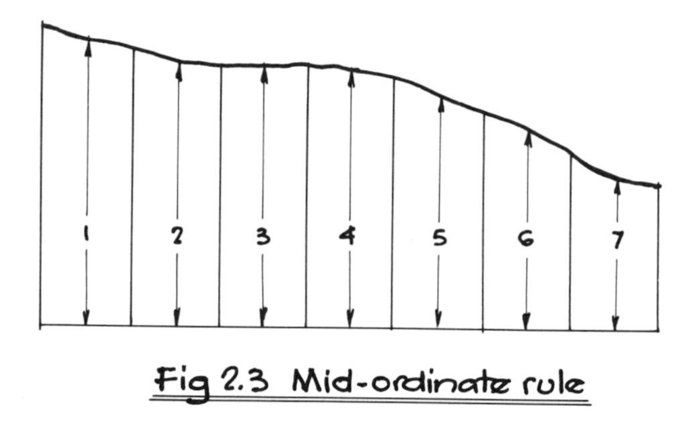

Fig 2.3 Mid-ordinate rule

Example 2.2
Determine the area of the lake considered earlier by using the mid-ordinate rule. For the figure shown, assume 12 equal-width strips of 5 m (even though one will be only 4 m, this is likely to give a reasonable answer). Now measure the mid-ordinates as in fig. 2.4. Area equals: 5 (14 + 17 + 21 + 26 + 32 + 36 + 36 + 23 + 18 + 11 + 8.5 + 5.5) = 5 x 248 = 1240 m^2. This is a slightly more accurate

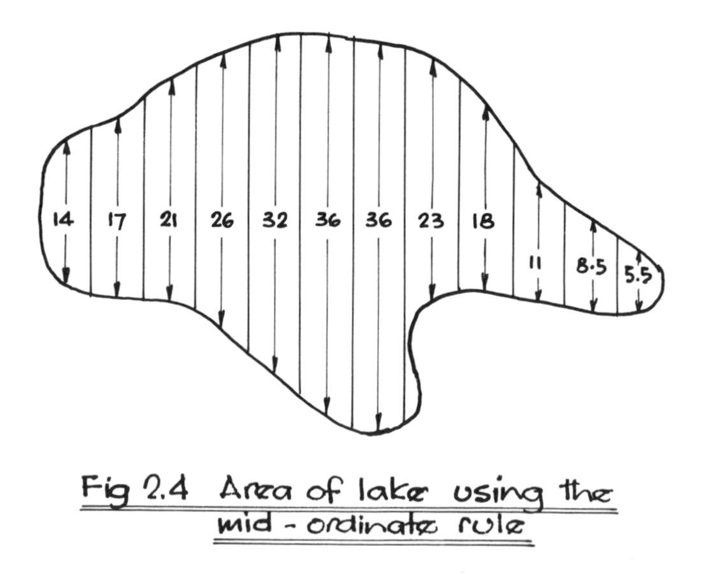

Fig 2.4 Area of lake using the mid-ordinate rule

answer than that obtained in example 2.1, and does not require the use of graph paper or the transfer of the shape from its original drawing.

TRAPEZOIDAL RULE

A rather similar method to that of the mid-ordinate rule is to consider each of the strips shown in fig. 2.3 as a trapezium with the irregular curve broken into a series of straight lines connecting the vertical lines of the strips (see fig. 2.5). We know, from Volume 1, that the area of a trapezium is half the sum of the parallel sides × the perpendicular distance between them, so that for fig. 2.5 this would be:

$$\left[\frac{a+b}{2}+\frac{b+c}{2}+\frac{c+d}{2}+\frac{d+e}{2}+\frac{e+f}{2}+\frac{f+g}{2}+\frac{g+h}{2}\right]\times w$$

This could be simplified to overcome the problem of dividing repeatedly by 2, to read:

$$\frac{w}{2}\left[a+2b+2c+2d+2e+2f+2g+h\right]$$

or $$\frac{w}{2}\left[a+h+2(b+c+d+e+f+g)\right]$$

which is expressed more easily as:

$$\frac{\text{width of strip}}{2}\left[\text{1st} + \text{last} + 2\,(\text{sum of internal strip heights})\right]$$

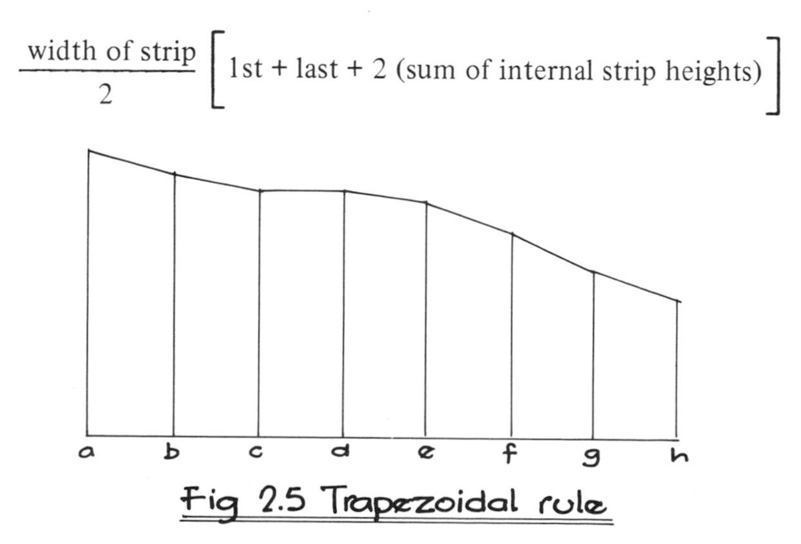

Fig 2.5 Trapezoidal rule

Obviously the greater the number of strips, the more accurate the answer is likely to be, since the straight lines will approximate more closely to the irregular curve.

Example 2.3

The shape, as illustrated in fig. 2.6, is divided into either four strips of 1.5 m width or twelve strips of 0.5 m width. Determine its area.

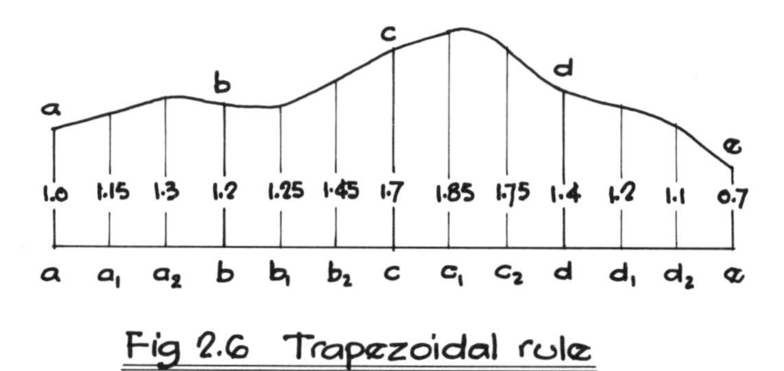

Fig 2.6 Trapezoidal rule

For four strips:

$$\text{area} = \frac{1.5}{2}\left[1 + 0.7 + 2\,(1.2 + 1.7 + 1.4)\right]$$

$$= 0.75 \times 10.3 = 7.725 \text{ m}^2$$

and from the figure we can see that although the lines connecting *ab* and *bc* probably compensate for each other, one being below the curve and the other above, the lines *cd* and *de* do not, since *both* are below the curve.

For twelve strips:

$$\text{area} = \frac{0.5}{2}\left[1 + 0.7 + 2\,(1.15 + 1.3 + 1.2 + 1.25 + 1.45 + 1.7 + 1.85 + 1.75 + 1.4 + 1.2 + 1.1)\right]$$

$$= 0.25 \times 32.4 = 8.1 \text{ m}^2$$

In this case the compensating errors are more evident so a more accurate approximation is achieved.

SIMPSON'S RULE

The most accurate of the approximate area methods for irregular shapes is Simpson's rule, which is a development of a precise method used for regular geometric shapes known as the *prismoidal rule*. Since the shapes dealt with, as stated at the beginning of the chapter, are invariably irregular, we need not dwell upon the prismoidal rule at this stage, but its derivative, Simpson's rule, states:

$$\text{area} = \frac{\text{width}}{3}\left[\text{1st} + \text{last} + 2 \times \text{sum of odds} + 4 \times \text{sum of evens}\right]$$

Note: the figure must *always* be divided into an *even* number of equal-width strips.

VOLUMES

So far we have been considering areas of irregular shapes, but in construction we are more usually concerned with *three-dimensional* rather than *two-dimensional* problems. How, then, may we apply the techniques covered so far? We know that a volume is an area x distance so, if we think of the ordinates (or strip lengths) being areas rather than linear measure, we can apply any of the three rules to calculate volumes. If considering a service trench with vertical sides, for example, we can consider the cross-sectional areas at regular intervals as ordinates in determining the volume of an excavation. If, however, we want to find the volume of an irregular-plan shape of irregular depth to excavation level, we would need to determine areas about one axis (using one of the rules) and then use these as ordinates about its other axis to determine volume. We would, of course, need to have established a grid to produce equal strip widths in each direction.

CASE STUDY 2.1: CALCULATION OF VOLUME OF EXCAVATION TO REDUCED LEVEL FOR A PILING CONTRACT

A construction site, of irregular level, slopes in two directions. It is intended to house a multi-storey building supported by piled foundations. A sub-contractor is appointed for the piling and as part of his requirements he asks that the site be levelled to the underside of pile caps. Figure 2.7 shows the plan of the proposed excavation with spot heights on a grid of 5 m squares. The spot heights and the reduced level are both in relation to the site datum,

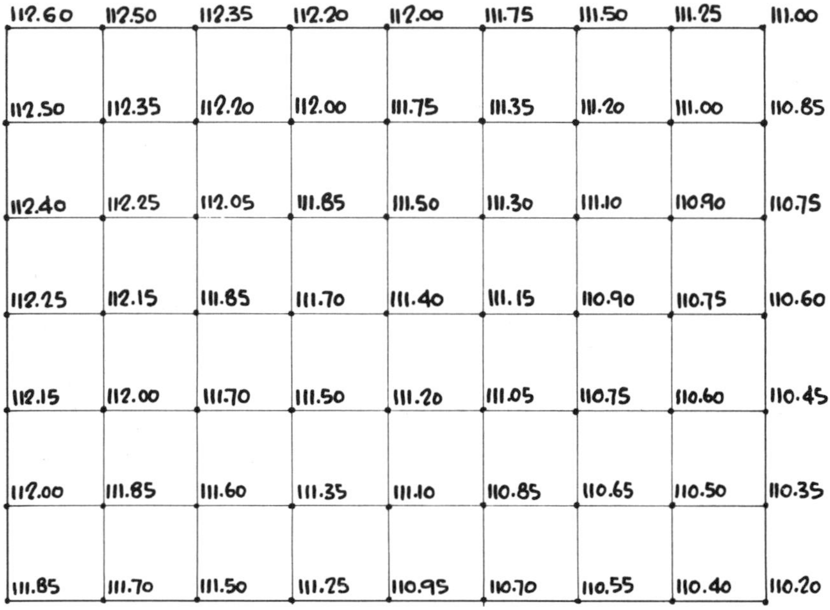

112.60	112.50	112.35	112.20	112.00	111.75	111.50	111.25	111.00
112.50	112.35	112.20	112.00	111.75	111.35	111.20	111.00	110.85
112.40	112.25	112.05	111.85	111.50	111.30	111.10	110.90	110.75
112.25	112.15	111.85	111.70	111.40	111.15	110.90	110.75	110.60
112.15	112.00	111.70	111.50	111.20	111.05	110.75	110.60	110.45
112.00	111.85	111.60	111.35	111.10	110.85	110.65	110.50	110.35
111.85	111.70	111.50	111.25	110.95	110.70	110.55	110.40	110.20

Fig 2.7 Construction Site 40 m x 30 m
to be excavated to reduced level of 108.00 m

so the depth of excavation is the difference of the two figures at each reading.

It can be assumed that the sides of the excavation are vertical, supported by temporary sheet piling, and the access ramp for the piling rig and excavating machinery may be discounted for the purpose of these calculations. Determine:

(a) the volume of earth excavated in reducing the level;
(b) the number of lorry loads to be carted away, given a bulking factor of 1.3 and a lorry capacity of 8.0 m^3;
(c) the cost of clearing the site, given excavation rate per hour, unbulked is 20 m^3 at a cost of £25.00 (including loading into lorries) and the cart-away cost per lorry load is £18.00.

Stage 1. Determine the cross-sectional areas, reading north to south, of the spot-height ordinates using Simpson's rule:

$$\text{Area 1} = \frac{5}{3}\left[4.60 + 3.85 + 2\,(4.40 + 4.15) + 4\,(4.50 + 4.25 + 4.00)\right]$$

$$= \frac{5}{3}\left[8.45 + 17.10 + 51.00\right] = \frac{5 \times 76.55}{3}$$

$$= 127.58 \text{ m}^2$$

$$\text{Area 2} = \frac{5}{3}\left[4.50 + 3.70 + 2\,(4.25 + 4.00) + 4\,(4.35 + 4.15 + 3.85)\right]$$

$$= \frac{5}{3}\left[8.20 + 16.50 + 49.40\right] = \frac{5 \times 74.10}{3}$$

$$= 123.50 \text{ m}^2$$

$$\text{Area 3} = \frac{5}{3}\left[4.35 + 3.50 + 2\,(4.05 + 3.70) + 4\,(4.20 + 3.85 + 3.60)\right]$$

$$= \frac{5}{3}\left[7.85 + 15.50 + 46.60\right] = \frac{5 \times 69.95}{3}$$

$$= 116.58 \text{ m}^2$$

$$\text{Area 4} = \frac{5}{3}\left[4.20 + 3.25 + 2\,(3.85 + 3.50) + 4\,(4.00 + 3.70 + 3.35)\right]$$

$$= \frac{5}{3}\left[7.45 + 14.70 + 44.20\right] = \frac{5 \times 66.35}{3}$$

$$= 110.58 \text{ m}^3$$

$$\text{Area 5} = \frac{5}{3}\left[4.00 + 2.95 + 2\,(3.50 + 3.20) + 4\,(3.75 + 3.40 + 3.10)\right]$$

$$= \frac{5}{3}\left[6.95 + 13.40 + 41.00\right] = \frac{5 \times 61.35}{3}$$

$$= 102.25 \text{ m}^2$$

$$\text{Area } 6 = \frac{5}{3}\left[3.75 + 2.70 + 2\,(3.30 + 3.05) + 4\,(3.35 + 3.15 + 2.85)\right]$$

$$= \frac{5}{3}\left[6.45 + 12.70 + 37.40\right] = \frac{5 \times 56.55}{3}$$

$$= 94.25 \text{ m}^2$$

$$\text{Area } 7 = \frac{5}{3}\left[3.50 + 2.55 + 2\,(3.10 + 2.75) + 4\,(3.20 + 2.90 + 2.65)\right]$$

$$= \frac{5}{3}\left[6.05 + 11.70 + 35.00\right] = \frac{5 \times 52.75}{3}$$

$$= 87.92 \text{ m}^2$$

$$\text{Area } 8 = \frac{5}{3}\left[3.25 + 2.40 + 2\,(2.90 + 2.60) + 4\,(3.00 + 2.75 + 2.50)\right]$$

$$= \frac{5}{3}\left[5.65 + 11.00 + 33.00\right] = \frac{5 \times 49.65}{3}$$

$$= 82.75 \text{ m}^2$$

$$\text{Area } 9 = \frac{5}{3}\left[3.00 + 2.20 + 2\,(2.75 + 2.45) + 4\,(2.85 + 2.60 + 2.35)\right]$$

$$= \frac{5}{3}\left[5.20 + 10.40 + 31.20\right] = \frac{5 \times 46.80}{3}$$

$$= 78.00 \text{ m}^2$$

Stage 2. Using the areas as ordinates, determine the volume, reading east to west, using Simpson's rule:

$$\text{Volume} = \frac{5}{3}\left[127.58 + 78.00 + 2\,(116.58 + 102.25 + 87.92) + 4\,(123.50 + 110.58 + 94.25 + 82.75)\right]$$

$$= \frac{5}{3}\left[205.58 + 613.50 + 1644.32\right] = \frac{5 \times 2463.40}{3}$$

$$= 4105.67 \text{ m}^3, \text{ say } 4106 \text{ m}^3$$

Answer to (a) = 4106 m^3

The calculation for lorry loads may be carried out directly, i.e. (volume x bulking factor)/lorry capacity. No. of lorries = (4106 x 1.3)/8 = 667.225, which must be rounded up.

Answer to (b) = 668 lorry loads

The cost calculations are two components based on the answers of (a) and (b), thus:

Excavation costs = (undisturbed volume x cost per hour)/volume rate per hour = (4106 x 25)/20 = £5132.50

Lorry costs = no. of lorries x cost per lorry = 668 x 18 = £12 024.00

Total cost = 5132.5 + 12 024.00 = £17 156.5, which would be rounded up to the nearest £50.

Answer to (c) = £17 200

From this it can be seen that cart-away costs contribute 70% of the excavation costs. This helps to explain why soil is stored for replacement or landscaping as a more economic alternative, in many instances, to carting away.

It should also be noticed from this example that much of the calculation process was repetitive. In such circumstances it can be beneficial to use a fairly simple computer routine, thus reducing the work load. *Note:* a programmable calculator may be used if it has sufficient memory (approximately 30 steps), but many can only cope with 20 to 24 steps. This is explained in chapter 8. Also, whichever method is used, be it calculator or computer, it is always worth carrying out an approximation to have confidence in the result. In the above example this can be achieved by taking the depth at the middle point (111.40 - 108.00) and multiplying by the plan area (40.00 x 30.00) to find the approximate volume of excavation, thus: 3.40 x 1200 = 4080 m^3. In this case the approximate figure is within 1% of the actual calculation but it is rare for such a close fit to be obtained.

So far we have only looked at volumes excavated to a level base and with vertical sides. In practice this is rarely the case. Sides of excavations are often 'cut to batters' using the soil's angle of repose to avoid the need for support work, though this is preferable for short-term excavation in dry-weather conditions only. Also the base to drainage trenches is usually 'cut to falls' to provide the necessary drainage. For large excavations the cross-section of 'battered cut' may be approximated to a triangle, but for the somewhat narrower trenches the full cross-section is best approximated to a Trapezium. In either case the width of cut at the top of the batter is dependent both on the angle of repose and the depth of cut (see fig. 2.8), the answer being found by using simple trigonometry, and it is sufficient to calculate the area by basic techniques prior to applying one of the rules outlined earlier in this chapter to determine the volume of cut.

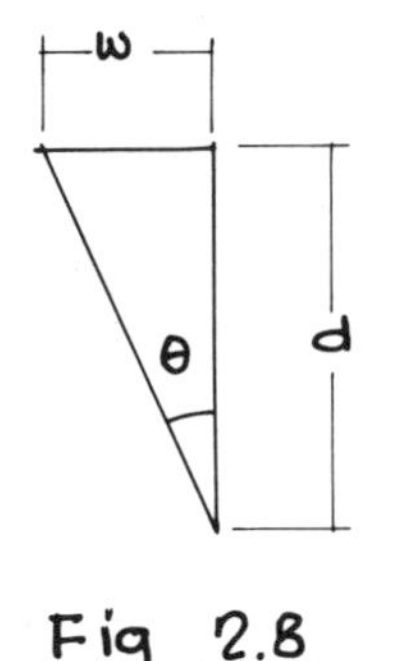

Fig 2.8

Example 2.4

A road is built to a gradient of 1 : 8 to form an underpass below an existing railway. The ground level is consistent and the slope to the cutting is to a batter of 1 : 4 (see fig. 2.9). Determine the volume of earth removed to form the embankment from the base level of the underpass to the existing ground level within the limits of 1 m below ground level and 3 m below ground level.

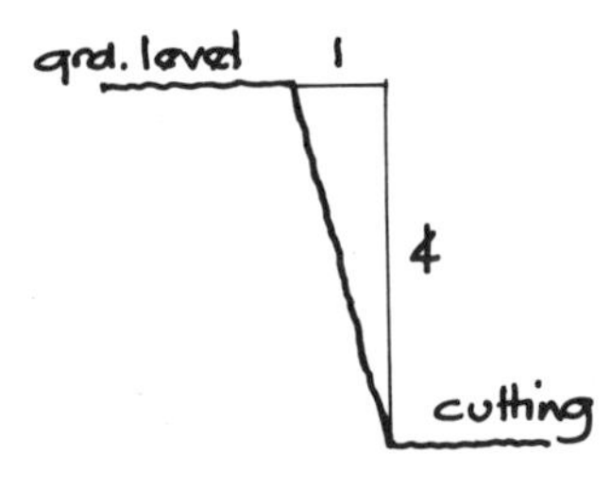

Fig 2.9

On this occasion we *can* use the *prismoidal rule* to which we referred earlier, since the volume follows the geometrical shape of a prism. The rule states:

$$V = \frac{l}{6} (A_F + 4A_M + A_L)$$

where l = length between first and last ordinates, A_F = area of first ordinate, A_M = area of middle ordinate, A_L = area of last ordinate.

However, in order to apply this rule, we must first determine all the values.

Stage 1. Find the *plan* length of the section of embankment knowing the distance fallen (2 m) and the gradient (1 : 8). Length = 2 x 8 = 16 m.

Stage 2. Find the depth of the middle ordinate and the width of cutting at each ordinate, knowing the batter (1 : 4), the first value for d (1.0 m) and the last value for d (3.0 m). Middle depth = 2.0 m, by interpolation. The ratios of the triangle shown in fig. 2.9 could be expressed as 0.25 : 1.0 rather than 1 : 4 by dividing through by four. Therefore:

w_F = 0.25 m
w_M = 0.50 m
w_L = 0.75 m

Stage 3. Calculate the areas of the three triangles, using ½ base x height, thus:

A_F = 0.125 m²
A_M = 0.500 m²
A_L = 1.125 m²

Stage 4. Apply the prismoidal rule to find the volume, thus:

$$\text{Volume} = \frac{16}{6}\left(0.125 + (4 \times 0.500) + 1.125\right)$$

$$= \frac{16 \times 3.25}{6} = 8.667 \text{ m}^3$$

Answer 8.67 m³ (to three significant figures)

Check by taking an approximation of middle area x length to satisfy yourself that the answer is of the right magnitude. Approximate volume = 0.5 x 16 = 8 m³.

Example 2.5

A drainage trench, of base width 600 mm and length of run 12 m, is excavated to a fall of 1 : 60. The ground is of irregular level and the sides of the excavation are trimmed to a batter of 1 : 3 to avoid the need for support. Given the datum of the trench base as 42.00 m at the start of its run and spot heights of the ground, at intervals of 2 m, as: 43.00, 43.10, 43.25, 43.20, 43.15, 43.05, 43.00, determine the volume of earth to be excavated.

In this case we cannot use the prismoidal rule, so the most accurate method is Simpson's rule. However, we must first determine the ordinate values, treating each as a trapezium.

Stage 1. Find the vertical height of each ordinate. To do this we need to know the base datum that corresponds to each spot height so that we can make a subtraction. Since the fall is 1 : 60, the total fall for 12 m is 200 mm and the incremental fall per 2 m is 33.3 mm. Thus the base datums are: 42.00, 41.97, 41.93, 41.90, 41.87, 41.83, 41.80, which produces ordinate heights of: 1.00, 1.13, 1.32, 1.30, 1.28, 1.22, 1.20.

Stage 2. Find the width of the trench at ground level for each ordinate. The width w for the triangle is $d/3$ because the batter is 1 : 3, so the overall width ($2w$ + base width) = $(2d/3) + b$, where b is constant at 600 mm. Thus the upper ordinate widths are:

(1) $\frac{2.00}{3} + 0.60 = 1.27$ (2) $\frac{2.26}{3} + 0.60 = 1.35$

(3) $\frac{2.64}{3} + 0.60 = 1.48$ (4) $\frac{2.60}{3} + 0.60 = 1.47$

(5) $\frac{2.56}{3} + 0.60 = 1.45$ (6) $\frac{2.44}{3} + 0.60 = 1.41$

(7) $\frac{2.40}{3} + 0.60 = 1.40$

Stage 3. Find the area of each trapezium ordinate, using ½ height x sum of top and bottom widths, thus:

Area 1 = $\frac{1.00}{2} \times 1.87 = 0.94$ Area 2 = $\frac{1.13}{2} \times 1.95 = 1.10$

Area 3 $= \frac{1.32}{2} \times 2.08 = 1.37$ Area 4 $= \frac{1.30}{2} \times 2.07 = 1.35$

Area 5 $= \frac{1.28}{2} \times 2.05 = 1.31$ Area 6 $= \frac{1.22}{2} \times 2.01 = 1.23$

Area 7 $= \frac{1.20}{2} \times 2.00 = 1.20$

Stage 4. Apply Simpson's rule, for strips of 2.00 m width =

$$\frac{2.00}{3}\left[0.94 + 1.20 + 2\,(1.37 + 1.31) + 4\,(1.10 + 1.35 + 1.23)\right]$$

$$\frac{2}{3}\ (2.14 + 5.36 + 14.72) = \frac{2 \times 22.22}{3}$$

$$= 14.81\ \text{m}^3$$

Check approximate width at middle 1.50 m and height 1.30 m giving area of $(1.30/2) \times 2.1 = 1.365\ \text{m}^2 \times$ length of 12 m = 16.38 m^3. The reason that this figure is high is because the ground rises and falls along the length of the trench.

From the example provided so far, you may be under the impression that the volume rules apply only to excavation work. Though the majority of applications are in this area of construction, the rules can also apply to constructional work itself, particularly when it follows a geometrical shape.

Example 2.6

A reinforced concrete chimney follows the form of a truncated cone having an outside diameter of 1.5 m at the top and 2.8 m at its base. To allow for the bending effect of the wind, the thickness of concrete is increased from 120 mm at the top to 300 mm at the base of the chimney. Given the vertical height of the chimney as 42 m, calculate the volume of concrete required for its construction.

We can use the prismoidal rule since the chimney is a 'frustrum', using the cross-sectional areas at top, middle and base.

Stage 1. Determine outside diameter and thickness of chimney wall at mid-height. Outside diameter = $(1.5 + 2.8) \div 2 = 2.15$ m. Thickness of wall = $(120 + 300) \div 2 = 210$ mm.

Stage 2. Calculate cross-sectional areas using:

$$\frac{\pi (D^2 - d^2)}{4}$$

where D = outside diameter, d = inside diameter.

$$\text{base} = \frac{\pi (2.80^2 - 2.20^2)}{4} = 2.36$$

$$\text{middle} = \frac{\pi (2.15^2 - 1.73^2)}{4} = 1.28$$

$$\text{top} = \frac{\pi (1.50^2 - 1.26^2)}{4} = 0.52$$

Stage 3. Calculate volume using:

$$\frac{42}{6} (2.36 + 4 \times 1.28 + 0.52)$$

$$= 7 \times 8 = 56 \text{ m}^3$$

Note: this example shows the value of reducing the thickness of concrete and/or the diameter as the height increases, since the *maximum* bending occurs at the base. For example, if the thickness were 300 mm throughout, the middle area would be: $[\pi (2.15^2 - 1.55^2)]/4 = 1.74$, and the top area would be: $[\pi (1.50^2 - 0.90^2)]/4 = 1.13$, giving a volume of:

$$\frac{42}{6} (2.36 + 4 \times 1.74 + 1.13) = 7 \times 10.45 = 73.15 \text{ m}^3,$$

an increase of 30%. If the cross-section were constant throughout, the volume would be $42 \times 2.36 = 99.12 \text{ m}^3$, an increase of 77%.

Exercises

2.1 Figure 2.10 shows a building plot with ordinates at 3 m intervals. Given a land cost of £280 000 per hectare, calculate the cost of the plot.

2.2 Figure 2.11 shows a longitudinal section through a service trench of constant width 400 mm. Determine the volume of excavated soil.

2.3 Figure 2.12 shows the spot heights of a building site. The dotted outline shows the area to be levelled to a datum of 75.00 m. Calculate the volume of soil to be removed.

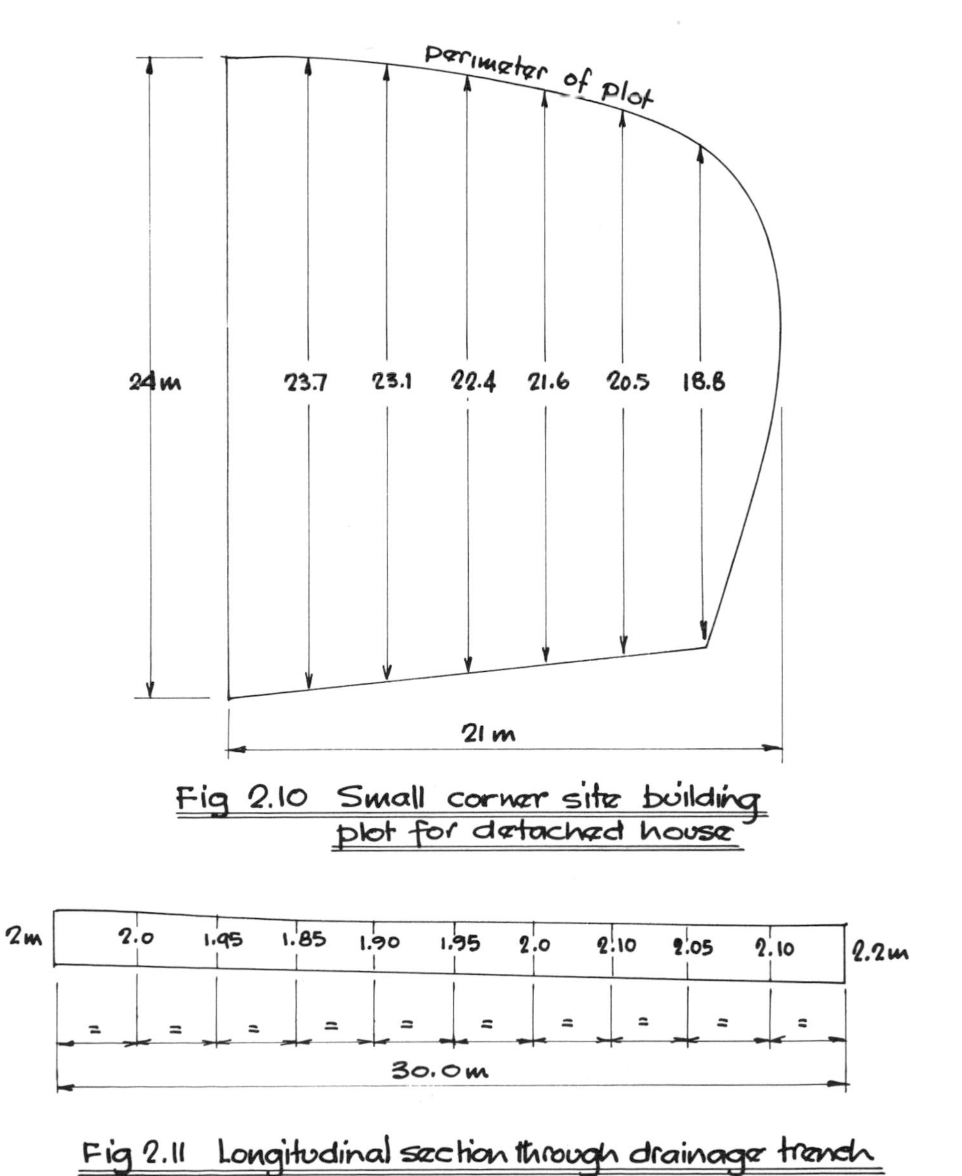

Fig 2.10 Small corner site building plot for detached house

Fig 2.11 Longitudinal section through drainage trench

77.4	77.4	77.5	77.55	77.6	77.7	77.75	77.8
77.5	77.5	77.6	77.65	77.7	77.8	77.85	77.9
77.5	77.65	77.7	77.75	77.8	77.85	77.9	78.0
77.65	77.7	77.8	77.85	77.9	77.95	78.0	78.1
77.75	77.8	77.85	77.9	77.95	78.05	78.1	78.2
77.85	77.9	77.95	78.0	78.1	78.15	78.2	78.3
77.9	78.0	78.1	78.15	78.2	78.25	78.3	78.3

Fig 2.12 Spot heights on 5m x 5m grid

2.4 A random rubble retaining wall forms one boundary to a playing field. Its section is variable, as shown in fig. 2.13, since the height of retained earth increases along its length. Knowing that the increase in height and base width are regular, calculate the volume of rubble required.

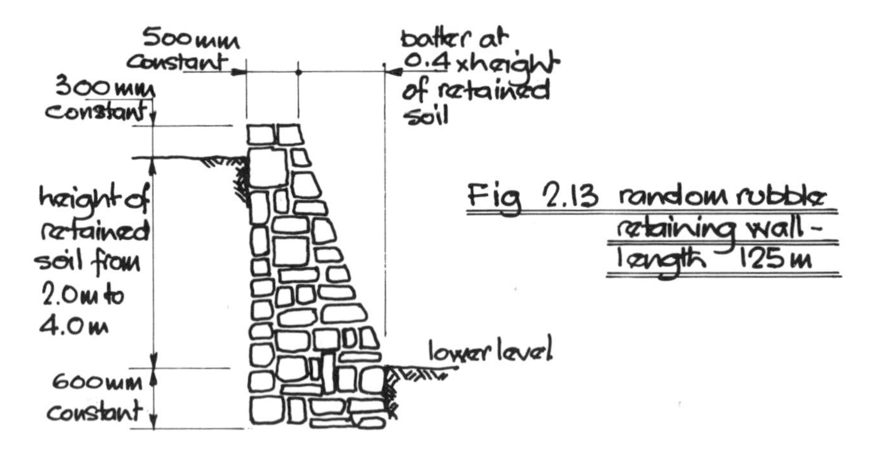

Fig 2.13 random rubble retaining wall - length 125m

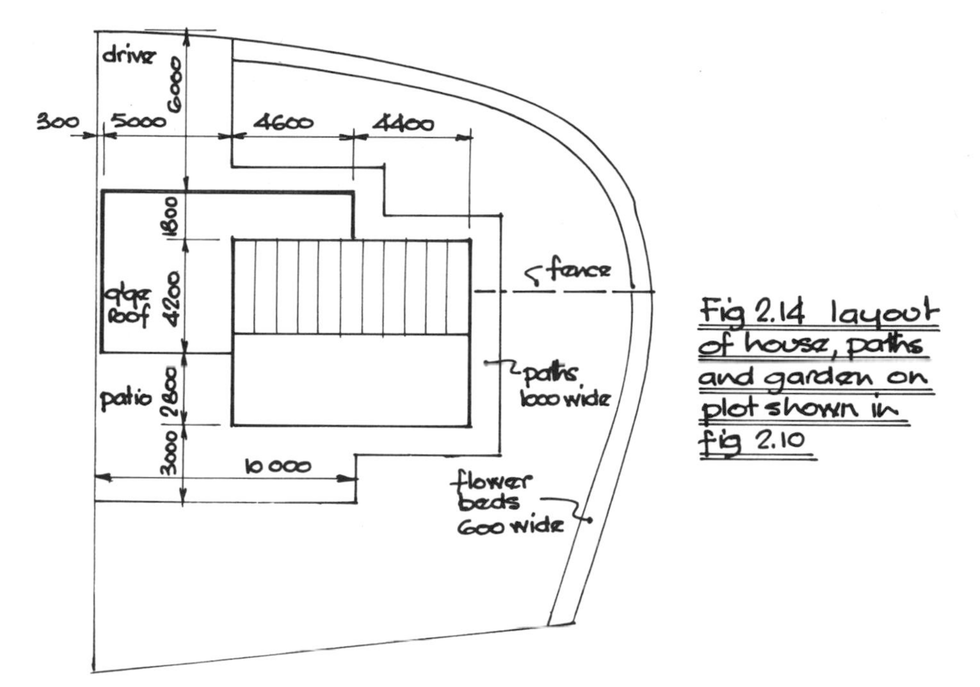

Fig 2.14 layout of house, paths and garden on plot shown in fig 2.10

2.5 Figure 2.14 shows a building plot with the house, paths and flowerbeds outlined. Determine, from the dimensions given, the area to be turfed.

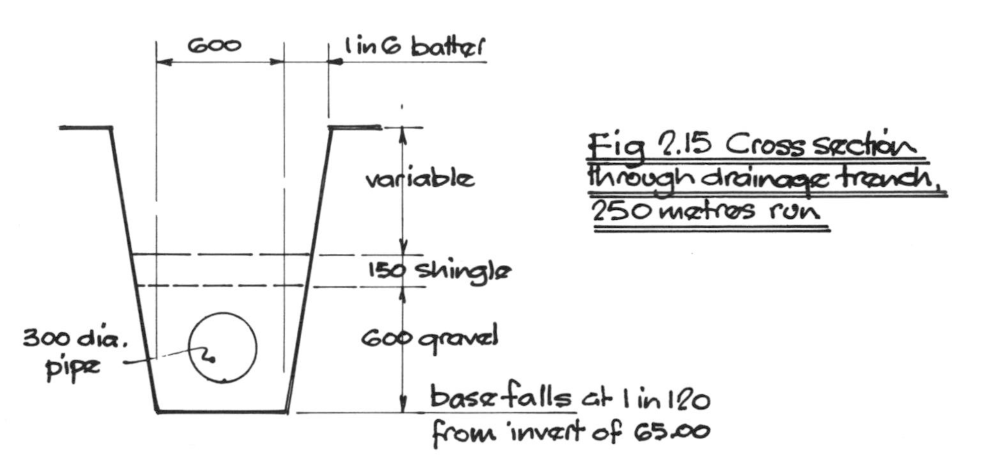

Fig 2.15 Cross section through drainage trench, 250 metres run

Record of spot heights at intervals of 10 metres along trench:

66.80 66.75 66.75 66.70 66.70 66.65 66.65 66.60 66.55
66.50 66.50 66.45 66.40 66.35 66.35 66.30 66.25 66.20
66.20 66.15 66.15 66.10 66.05 66.05 66.00 65.95

Assignment 2.1

Figure 2.15 shows the section through a drainage trench, indicating the pipe size and dimensions of granular backfill together with spot heights and trench-base datums. A large proportion of the excavated earth is to be retained for backfilling and consolidating the excavation, and the remainder is to be carted away. Establish current prices for excavation and backfilling of the earth and for carting away the surplus (bulking factor of 1.25), the cost of granular backfill per cubic metre and pipework per metre run. From this determine the cost of excavating, drain laying and making good the site.

3 First and Second Moments of Area

FIRST MOMENTS OF AREA

In *Construction Science* Volume 2 we used moments of force to determine the centre of gravity of a system of forces. In much the same way we can use moments of area (or area moments) to determine the centroid of a system of shapes. This technique relies on us knowing the centroid of each individual shape (see chapter 8, *Construction Mathematics* Volume 1) so that we can take moments about an arbitrary point of each individual area and summate these to find the total area moment. From this, knowing the total area, we can locate its centroid.

Example 3.1
Figure 3.1 shows a composite of three rectangles. In order to find its centroid we need to take area moments about two axes (for convenience's sake, the vertical and horizontal planes) since one axis will only indicate a line on which it lies. The positions from which we take moments are unimportant so long as we take all moments in *one* plane from *one* position. It may be located inside or outside the shape and sometimes it is quite useful to choose the

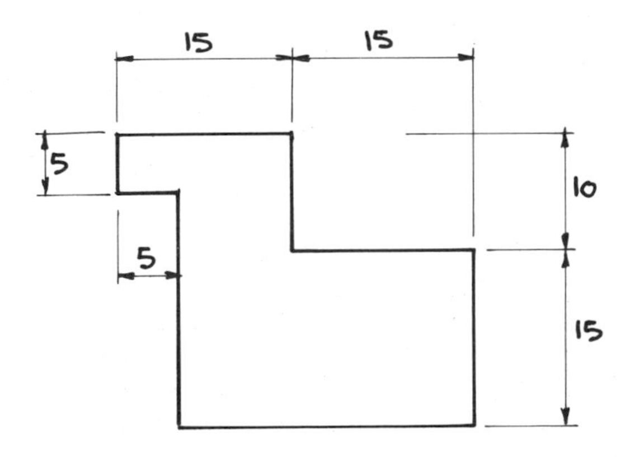

Fig 3.1 Composite rectilinear figure

centroid of one of the component shapes to minimise the calculation, since, having no distance, that shape cannot have a moment of area. It is also unimportant in relation to how we subdivide the complex shape. Figure 3.2 shows three alternative ways of approaching the problem. To prove this we will choose two alternatives for this example and you can try the one remaining.

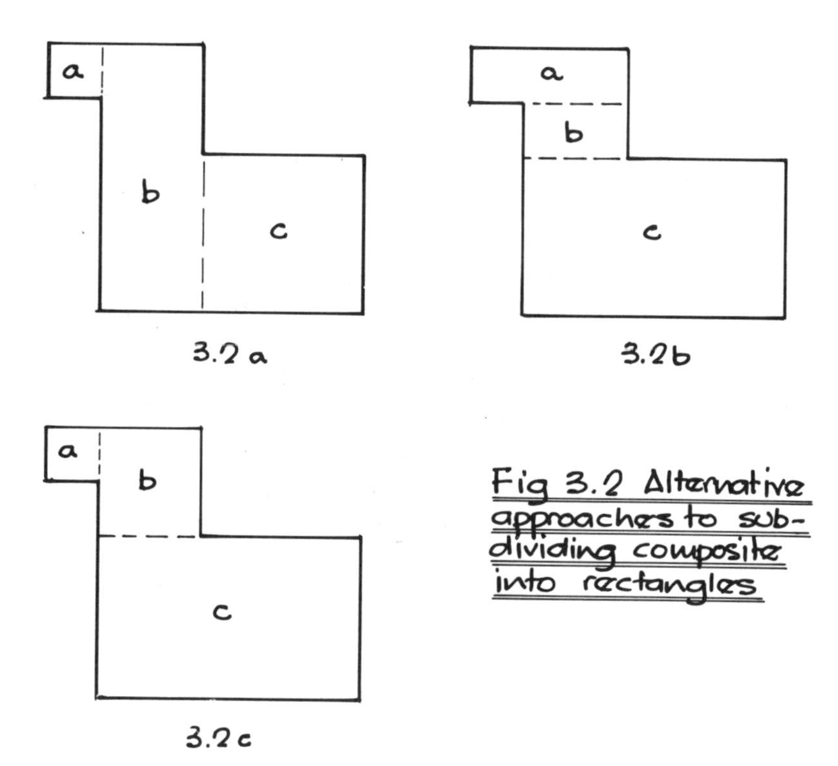

Fig 3.2 Alternative approaches to sub-dividing composite into rectangles

1st alternative – fig. 3.2a. Vertical plane, taking a position 10 mm to the left of the point X: the centroids of areas *a, b* and *c* are, respectively: 12.5 mm, 20 mm and 32.5 mm from the axis (see fig. 3.3v). The areas are, respectively, 25 mm^2, 250 mm^2 and 225 mm^2, which, when summated, produce a total area of 500 mm^2. Since there is a fair amount of repetitive calculation, it can simplify matters if we adopt a 'tabulated' approach similar to that shown in *Construction Science* Volume 2, when dealing with retaining walls. This ensures that we remember to consider each component shape when carrying out the calculations, and reduces the risk of error, thus:

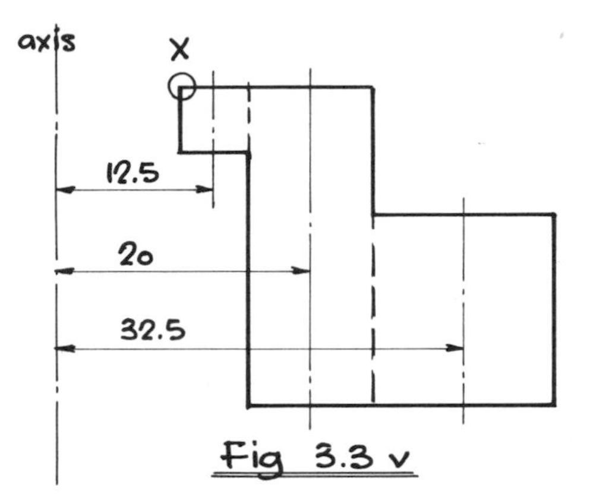

Fig 3.3 v

	Area of unit shape	Distance of centroid from vertical axis	Area moment produced
a	25	12.5	312.5
b	250	20.0	5 000.0
c	225	32.5	7 312.5
Σ	500	$\bar{x}$	12 625.0

We can see that 500 x $\bar{x}$ = 12 625, therefore $\bar{x}$ = 12 625 ÷ 500 = 25.25, or 15.25 mm to the right of X.

Horizontal plane, taking a position along the top of the composite shape: the centroids of areas a, b and c are, respectively: 2.5 mm, 12.5 mm, 17.5 mm (see fig. 3.3h).

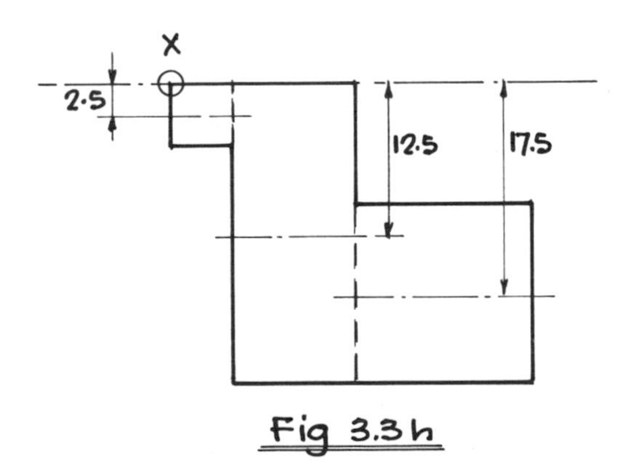

Fig 3.3 h

	Area of unit shape	Distance of centroid from horizontal axis	Area moment produced
a	25	2.5	62.5
b	250	12.5	3 125.0
c	225	17.5	3 937.5
Σ	500	$\bar{y}$	7 125.0

$\bar{y}$ = 7125 ÷ 500 = 14.25 mm below X. Centroid is shown in fig. 3.4.

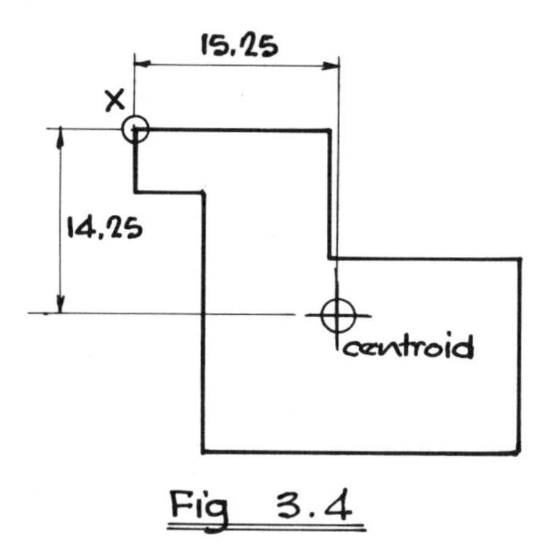

Fig 3.4

2nd alternative – fig. 3.2b. This time we will operate from the centroid of shape a and proceed directly to the tabulation, thus:

	Area of unit shape	Distance of centroid from vertical axis	Area moment produced
a	75	0	0
b	50	2.5	125
c	375	10.0	3 750
Σ	500	$\bar{x}$	3 875

$\bar{x}$ = 3875 ÷ 500 = 7.75 or 15.25 to right of X.

	Area of unit shape	Distance of centroid from horizontal axis	Area moment produced
a	75	0	0
b	50	5	250
c	375	15	5 625
Σ	500	$\bar{y}$	5 875

$\bar{y} = 5875 \div 500 = 11.75$ or 14.25 below X.

These coincide exactly with alternative 1, showing that both are accurate. You will notice from this example that, since the areas are constant, we *could* tabulate both axes concurrently if we wished to do so.

You may wonder as to the value of this in terms of construction. If you have read *Construction Science* Volume 2, you will already have seen its relevance to retaining walls, and if you go on to study structural analysis, the use of area moments is valuable in determining beam deflection and fixed-end moments to restrained beams. The most immediate value, however, is in determining the positions of neutral axes for beams of asymmetrical cross-section.

Example 3.2

Determine the location of the neutral axes in the major planes (X–X and Y–Y) for the beam section shown in fig. 3.5

Stage 1. Divide the beam into three rectangles (a, b, c) and take area moments in both planes as indicated in figs 3.6a–c. To simplify

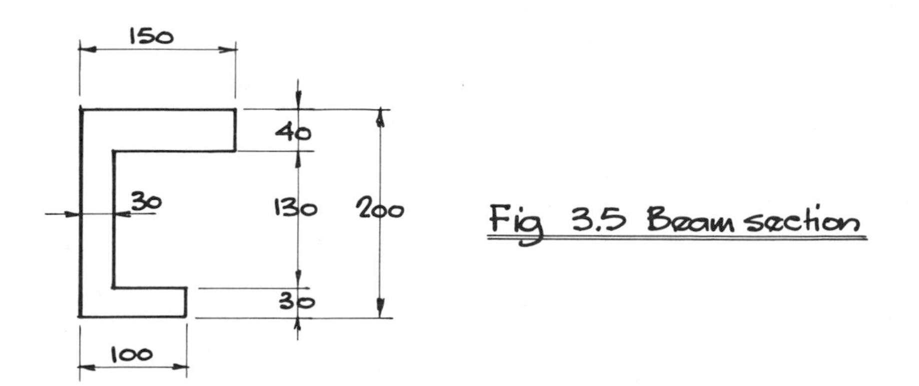

Fig 3.5 Beam section

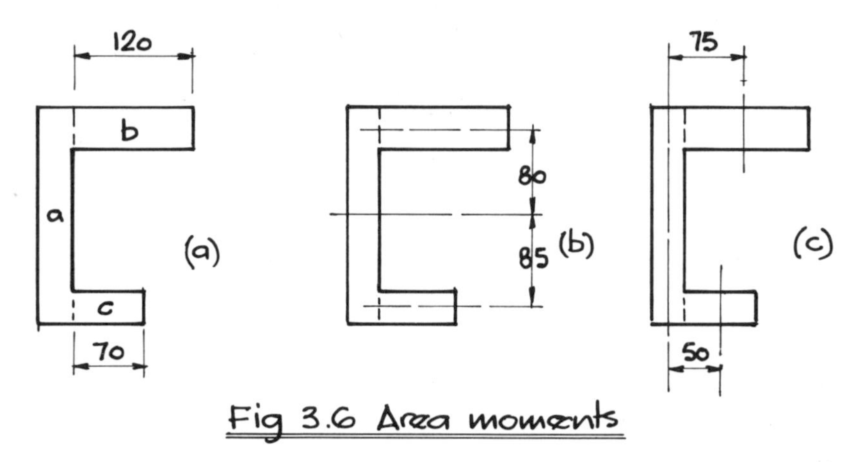

Fig 3.6 Area moments

the calculations we will take moments through the centroids of area *a* for each plane using a composite tabulation.

Note: It is important when taking moments from within the section to take account of opposing moments. In this example, for instance, the area moments on the X–X axis rotate in *opposite* directions. Thus area moment *c* will be negative.

	Area of unit shape	Distance of centroid X–X	Area moment X–X plane	Distance of centroid Y–Y	Area moment Y–Y plane
a	6 000	0	0	0	0
b	4 800	80	384 000	75	360 000
c	2 100	-85	-178 500	50	105 000
Σ	12 900	$\bar{x}$	205 500	$\bar{y}$	465 000

$\bar{x}$ = 205 500 ÷ 12 900 = 15.93 mm from centroid of *a*, or 84.07 mm from the top of the section.

$\bar{y}$ = 465 000 ÷ 12 900 = 36.05 mm from centroid of *a*, or 51.05 mm from the left-hand side of the section (see fig. 3.7).

The shape chosen for the example was asymmetrical about both the X–X and Y–Y axes so that the procedures could be illustrated fully. It is, however, more usual to provide beams that are asymmetrical about the X–X axis only, so that the Y–Y axis passes directly through the middle of the section. The imbalance may be in either the breadth or thickness of flanges or a combination of both, and it is often the top flange which is larger to improve its resistance to buckling when the section acts in bending.

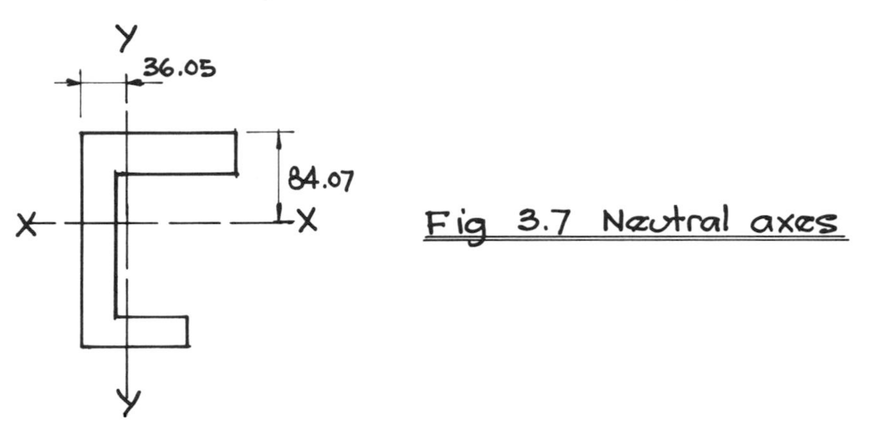

Fig 3.7 Neutral axes

SECOND MOMENTS OF AREA

Having located the neutral axes it is now possible for us to develop several mathematical properties for any cross-section, the majority of which are derived from the section's *second moment of area* (or moment of inertia, I) about either axis.

If we were to divide the section into a very large number of very thin strips, we could, by taking second moment areas (areas x distance squared), determine the gross second moment of area for either axis quite accurately, and in chapter 5 on integral calculus we shall do this. It is, however, much more convenient to consider the fairly large areas that we use in locating the neutral axes, and to add their individual I values to the effects of their location with regard to the neutral axes.

Thus, for either axis, we can use the equation:

$$I_G = \Sigma I_C + ak^2$$

where I_G = the gross moment of inertia for a plane; I_C = the component I values in that plane for each area; a = each area; and k = the distance from the centroid of each area to the neutral axis for that plane.

Example 3.3

Determine the second moment of area about the X–X and Y–Y axes for the cross-section shown in fig. 3.8.

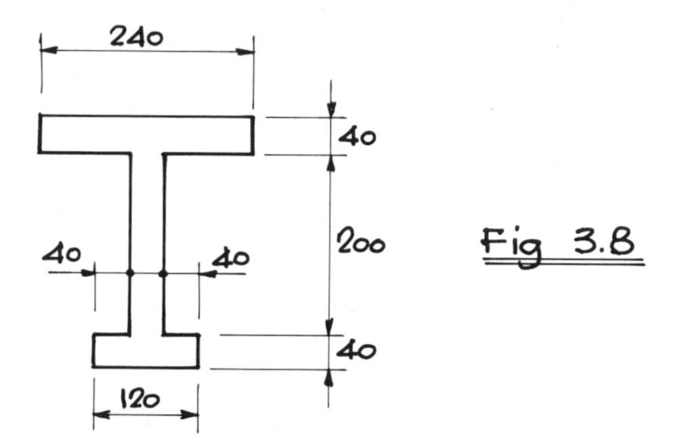

Y–Y axis passes through the centre of the section and therefore the centre of each area. It will be straightforward to calculate the I_{Y-Y} since there are *no* ak^2 components. The X–X axis, by inspection, will be located nearer the top flange than the bottom and must be determined. This in turn will affect the I_{X-X} value.

Stage 1. Locate the neutral axis (NA) position for the X–X plane by area moments. We can take moments through the centroid of area a (see fig. 3.9).

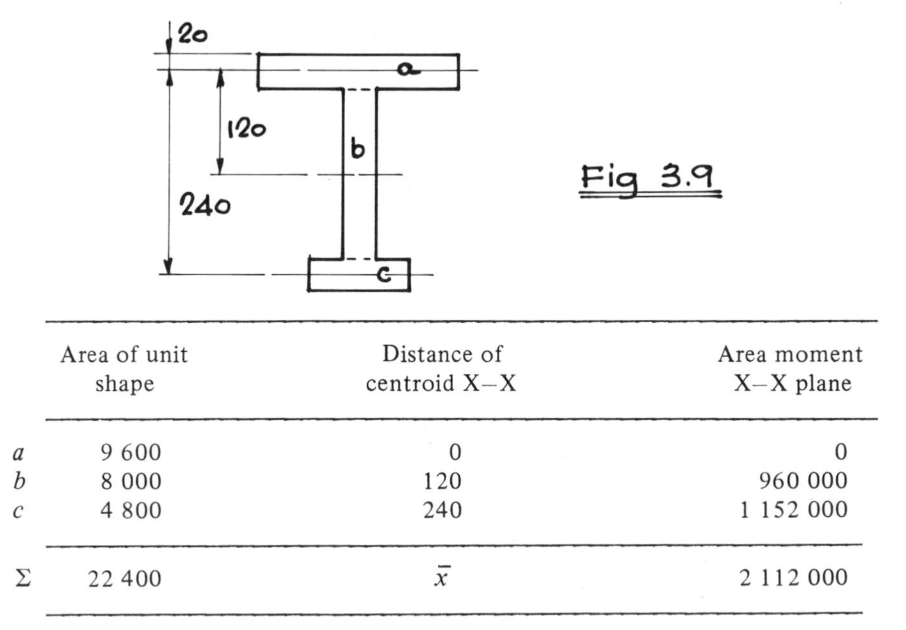

	Area of unit shape	Distance of centroid X–X	Area moment X–X plane
a	9 600	0	0
b	8 000	120	960 000
c	4 800	240	1 152 000
Σ	22 400	$\bar{x}$	2 112 000

$\bar{x}$ = 2 112 000 ÷ 22 400 = 94.29 mm from centroid of a, or 114.29 mm from limit of top flange.

Stage 2. Calculate value of I_{X-X} by taking second moments about the neutral axis (see fig. 3.10).

$$I_G = \Sigma I_C + ak^2$$

where I for a rectangle $= bd^3/12$.

$$I_G = \Sigma \frac{240 \times 40^3 + 40 \times 200^3 + 120 \times 40^3}{12}$$

$$+ 240 \times 40 \times 94.29^2 + 40 \times 200 \times 25.71^2 + 120 \times 40 \times 145.71^2$$

Although it is quite possible to calculate these numbers as they stand, it is more logical to introduce index powers to simplify the figures. Since I values are commonly expressed in $mm^4 \times 10^6$ for structural calculations, this can be introduced *before* making the arithmetical calculations. If we think in terms of hundreds (10^2) for the *depth* and *distance*, this will produce the required effect, i.e. $[b \times (d \times 10^2)^3]/12$ and $b \times (d \times 10^2) \times (k \times 10^2)^2$ will become $(b \times d^3 \times 10^6)/12$ and $b \times d \times k^2 \times 10^6$. This reduces the equation to:

$$I_G = 10^6 \; \Sigma \frac{240 \times 0.4^3 + 40 \times 2^3 + 120 \times 0.4^3}{12} +$$

$$240 \times 0.4 \times 0.9429^2 + 40 \times 2 \times 0.2571^2 + 120 \times 0.4 \times 1.4571^2$$

$$= 10^6 \; \Sigma \; 28.59 + 85.35 + 5.29 + 101.91$$

$$I_{X-X} = 221.14 \times 10^6 \text{ mm}^4$$

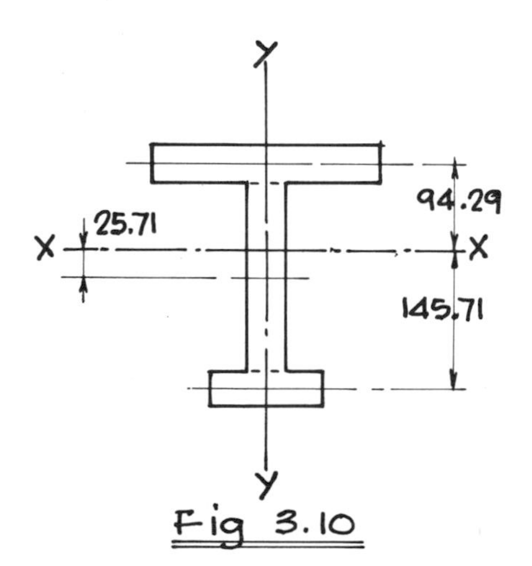

Fig 3.10

Obviously, with such complex calculations it is *essential* that you have an approximate idea of the end figure. This can be obtained quite quickly, e.g.

$$\begin{aligned}
250 \times 0.06 &= 15\\
+\ 40 \times 8 &= 320\\
+\ 125 \times 0.06 &= 7.5\\
\hline
&\ 342.5 \div 12 \simeq 30\\
+\ 90 \times 0.8 &\simeq 70\\
+\ 80 \times 0.06 &\simeq 50\\
+\ 50 \times 2 &= 100\\
\hline
&\ 220 \rightarrow \frac{220}{\ } \\
&\simeq 250 \times 10^6
\end{aligned}$$

which is within 15% of the correct answer.

Stage 3. Calculate the value of I_{Y-Y}.
Since the axis passes through the centre of the cross-section, there are no second moments of area additional to those of the areas themselves so $I_{Y-Y} = \Sigma\, I_C$; which, using indices, equals:

$$10^6\ \Sigma\ \frac{40 \times 2.4^3 + 200 \times 0.4^3 + 40 \times 1.2^3}{12}$$

$$= 10^6\ \Sigma\ \frac{552.96 + 12.8 + 69.12}{12} = \frac{634.88 \times 10^6}{12}$$

$$= 52.91 \times 10^6\ \text{mm}^4$$

Checking by approximation:

$$\begin{aligned}
40 \times 12.5 &= 500\\
+\ 200 \times 0.06 &= 12\\
+\ 40 \times 2 &= 80\\
\hline
&\ 592 \div 12 \simeq 50 \times 10^6
\end{aligned}$$

This is within 5% of the correct answer.

The reason that an index coefficient of 10^6 is used is so that the final figures are of a manageable size. When using steelwork sections which, because of the material's strength, are somewhat slimmer, it is more sensible to use 10^4. In part this is to retain manageable figures, but a second consideration is that the standard section properties are expressed in cm^4 (which, as you know, is *not* an SI value), which is the same as $mm^4 \times 10^4$. Similarly, other properties are in centimetre units. Conversely, concrete sections can be quite bulky by comparison, and I values may be expressed in terms of $mm^4 \times 10^8$ if so desired. The important factor is what we do with such values and that we remember the index coefficients that we have used.

OTHER MATHEMATICAL PROPERTIES OF STRUCTURAL SECTIONS

Earlier in the chapter we talked of associated mathematical properties of particular interest to students of structural mechanics. These are referred to in *Construction Science* Volume 2, but it is well worth restating them here since students do not always understand the interdependence of science, mathematics and construction.

Section Modulus (Z)

This is *not* to be confused with the first moment of area, even though the units are to a similar power, i.e. mm^3. The basic equation for the section modulus is: $Z = I/y$, where I = second moment of area for axis considered, and y = distance from neutral axis (NA) to extreme fibres of the section. When the section is asymmetrical, this will result in *two* Z values for that axis (or those axes) since there are two distances from the neutral axis. For the section considered in example 3.3, for instance, there would be *two* Z_{X-X} values: $(221.14 \times 10^6)/114.29 = 1935 \times 10^3$ mm^3, and $(221.14 \times 10^6)/165.71 = 1335 \times 10^3$ mm^3, but only *one* Z_{Y-Y} value: $(52.91 \times 10^6)/120 = 441 \times 10^3$ mm^3.

Radius of gyration (r_{X-X} or r_{Y-Y})

This is simply a value obtained by dividing the second moment of area by the area itself and then square rooting the result, i.e.

$$r_{X-X} = \sqrt{\frac{I_{X-X}}{A}} \qquad r_{Y-Y} = \sqrt{\frac{I_{Y-Y}}{A}}$$

which, for example 3.3, would produce values of:

$$r_{X-X} = \sqrt{\frac{221.14 \times 10^6}{2.24 \times 10^4}} = 99.36 \text{ mm}$$

$$r_{Y-Y} = \sqrt{\frac{52.91 \times 10^6}{2.24 \times 10^4}} = 48.60 \text{ mm}$$

The usefulness of these properties can be seen in the design of beams and columns. For beams their application is:

I_{X-X}	to establish deflection
Z_{X-X}	to establish resistance to bending
r_{Y-Y}	to establish resistance to buckling for the compression flange (or zone)

For columns their application is:

I_{X-X} and I_{Y-Y}	the 'stiffness' of the column
Z_{X-X} and Z_{Y-Y}	to establish contribution of the column to resist bending caused by eccentric loading
r_{X-X} and r_{Y-Y}	to establish resistance to buckling about either axis

This does not mean that, for every beam design, we need to use *all* the properties, or that, for every column design, bending will occur. The values are for use as and when appropriate to a particular set of circumstances.

CASE STUDY 3.1

A manufacturer of standard timber beam sections supplies these for use in flooring and flat-roof construction. Figure 3.11 shows a built-up beam in cross-section; its flanges are softwood and its web plywood. Ply web stiffeners are provided to eliminate buckling problems. Given the following information:

Allowable bending stress	=	12 N/mm^2
Young's modulus (E)	=	10 000 N/mm^2
Maximum allowable deflection	=	0.003 of span
BM (max.)	=	$WL/8$
Deflection (max.)	=	$5WL^3/384EI$

produce safe load tables for spans of 3, 4 and 5 m with joist spacings of 300, 400 and 600 mm.

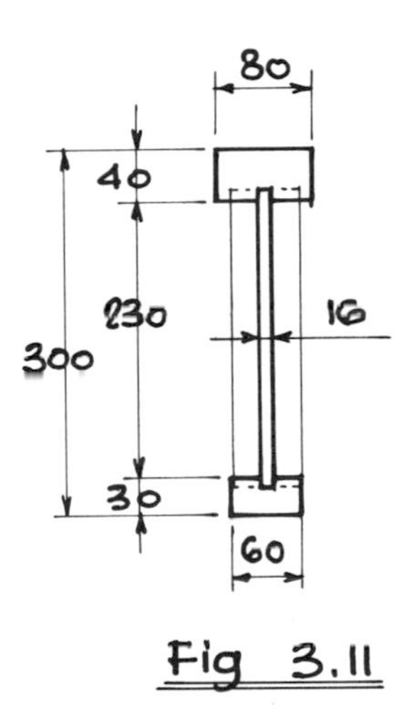

Fig 3.11

Stage 1. Determine location of the neutral axis in the X–X plane by taking area moments from the top of the beam section. Although the ply extends into the softwood, we can still assume three rectangles, 80 x 40, 16 x 230 and 60 x 30.

	Area of unit shape	Distance of centroid X–X	Area moment X–X plane
a	3 200	20	64 000
b	3 680	155	570 400
c	1 800	285	513 000
Σ	8 680	$\bar{x}$	1 147 400

$\bar{x} = 1\,147\,400 \div 8\,680 = 132.2$ mm

Stage 2. Determine I_{X-X} value of section, using $I_G = \Sigma\, I_C + ak^2$

$$I_{X-X} = 10^6\ \Sigma\ \frac{80 \times 0.4^3 + 16 \times 2.3^3 + 60 \times 0.3^3}{12}$$

$$+\ 80 \times 0.4 \times 1.122^2 + 16 \times 2.3 \times 0.228^2 + 60 \times 0.3 \times 1.528^2$$

$$= 10^6\ \Sigma\ \frac{5.12 + 194.67 + 1.62}{12} + 40.28 + 1.91 + 42.03$$

$$= 10^6\ \Sigma\ \frac{201.41}{12} + 84.22 = 101 \times 10^6\ \text{mm}^4$$

Stage 3. Determine Z_{X-X} values, using $Z = I/y$.

$$Z_{X-X}\ \text{(top flange)} = \frac{101 \times 10^6}{132.2} + 764 \times 10^3\ \text{mm}^3$$

$$Z_{X-X}\ \text{(bottom flange)} = \frac{101 \times 10^6}{167.8} + 602 \times 10^3\ \text{mm}^3$$

For safe load calculations we must use the lower of the two values, i.e. $Z_{X-X} = 602 \times 10^3\ \text{mm}^3$.

Stage 4. Determine the resistance moment of the beam, MR = fZ (see *Construction Science* Volume 2).

$$\text{MR} = 12 \times 602 \times 10^3\ \text{N mm}$$

But since BM values (bending moments) are normally expressed in kilonewton metres, we must divide by 10^6. Thus

$$\text{MR} = \frac{12 \times 602 \times 10^3}{10^6} = 7.224\ \text{kN m}$$

Stage 5. Determine deflection limits for spans = 0.003 l

3 m = 9 mm
4 m = 12 mm
5 m = 15 mm

Stage 6. Determine loading rates for the nine conditions in terms of w kN/m² $W = w \times \text{spacing} \times \text{span}$ and is best tabulated (values for W in kN):

Spacing	Span		
	3 m	4 m	5 m
300	0.9 w	1.2 w	1.5 w
400	1.2 w	1.6 w	2.0 w
600	1.8 w	2.4 w	3.0 w

Stage 7. Determine bending moment values for the nine conditions using $WL/8$ where W is kilonewtons and L is in metres, giving kilonewton metre values. Tabulate as before.

Spacing	Span		
	3 m	4 m	5 m
300	0.3375 w	0.6 w	0.9375 w
400	0.45 w	0.8 w	1.25 w
600	0.675 w	1.2 w	1.875 w

Stage 8. Determine deflection in terms of w, using $5WL^3/384EI$. Since the deflection will be in millimetres and the E value is in newtons per square metre, this means we must convert W from kilonewtons to newtons ($\times 10^3$) and L^3 from cubic metres to cubic millimetres ($\times 10^9$) giving the equation

$$\frac{5 \times W \times 10^3 \times L^3 \times 10^9}{384 \times 10\,000 \times 101 \times 10^6}$$

Simplifying: $$\frac{5 \times WL^3 \times 10^{12}}{384 \times 1.01 \times 10^{12}} = WL^3 \times 1.29 \times 10^{-2}$$

For the values of W and L we can again tabulate, thus:

Spacing	Span		
	3 m	4 m	5 m
300	0.313 w	0.991 w	2.419 w
400	0.418 w	1.321 w	3.225 w
600	0.627 w	1.981 w	4.838 w

Stage 9. For each condition we now equate w in terms of bending and deflection to find the *lesser* value since this will govern the safe load, i.e. it is no use satisfying bending if the deflection is excessive, or satisfying deflection limits if the beam fails in bending.

3 m span:

300 mm spacing (bending)	$0.3375\,w = 7.224$	
Therefore	$w = 7.224/0.3375$	$= 21.4\ \text{kN/m}^2$
300 mm spacing (deflection)	$0.313\,w = 9$	
Therefore	$w = 9/0.313$	$= 28.8\ \text{kN/m}^2$
400 mm spacing (bending)	$0.45\,w = 7.224$	
	$w = 7.224/0.45$	$= 16.05\ \text{kN/m}^2$
400 mm spacing (deflection)	$0.418\,w = 9$	$w = 21.53\ \text{kN/m}^2$
600 mm spacing (bending)	$0.675\,w = 7.224$	$w = 10.70\ \text{kN/m}^2$
600 mm spacing (deflection)	$0.627\,w = 9$	$w = 14.35\ \text{kN/m}^2$

Bending governs for each case.

4 m span:

300 mm spacing (bending)	$0.6\,w = 7.224$	$w = 12.04\ \text{kN/m}^2$
300 mm spacing (deflection)	$0.991\,w = 12$	$w = 12.11\ \text{kN/m}^2$
400 mm spacing (bending)	$0.08\,w = 7.224$	$w = 9.03\ \text{kN/m}^2$
400 mm spacing (deflection)	$1.321\,w = 12$	$w = 9.08\ \text{kN/m}^2$
600 mm spacing (bending)	$1.2\,w = 7.224$	$w = 6.02\ \text{kN/m}^2$
600 mm spacing (deflection)	$1.981\,w = 12$	$w = 6.06\ \text{kN/m}^2$

Again bending governs, but only marginally.

5 m span:

300 mm spacing (bending)	$0.9375\,w = 7.224$	$w = 7.70\ \text{kN/m}^2$
300 mm spacing (deflection)	$2.419\,w = 15$	$w = 6.20\ \text{kN/m}^2$
400 mm spacing (bending)	$1.25\,w = 7.224$	$w = 5.78\ \text{kN/m}^2$
400 mm spacing (deflection)	$3.225\,w = 15$	$w = 4.65\ \text{kN/m}^2$
600 mm spacing (bending)	$1.875\,w = 7.224$	$w = 3.85\ \text{kN/m}^2$
600 mm spacing (deflection)	$4.838\,w = 15$	$w = 3.10\ \text{kN/m}^2$

This time deflection governs in each case, showing that as the span increases the deflection becomes more critical.

Stage 10. Tabulate safe loads in kilonewtons per square metre.

Spacing	Span		
	3 m	4 m	5 m
300	21.40	12.04	6.20
400	16.05	9.03	4.65
600	10.70	6.02	3.10

This serves to show that although, when looked at as a whole, the calculations appear complex and rather unnerving, when they are broken down into ten stages, they become much more straightforward and easy to follow.

From a construction point of view it can be seen that if the design load for all three spans were 6.00 kN/m^2, the spacing for both 3 and 4 m would be 600 mm but for 5 m it would need to be 300 mm. The reason for restricting the spacings to these three alternatives is: (a) boarding (either roof or floor) is in 2.4 m lengths, (b) 600 mm is the maximum span that can be safely accommodated by the boarding.

CASE STUDY 3.2

A manufacturer of built-up box-section timbers supplies them for use in stud walling. Figure 3.12 shows the cross-section of one of these studs, comprising softwood blocking and ply webs. The webs, however, are non-continuous to allow for the passage of service runs, such as hot and cold water and central heating, and do not contribute fully to the sectional properties.

Given the following information:

Centres of studs 600 mm
Effective height (major axis) 2400 mm
Load sharing factor 1.1
Basic compressive stress 5.5 N/mm^2 (Grade 50)
Medium-term loading for CP112, table 15

determine:

(a) the centres of cross studding required to the minor axis.
(b) the safe axial load per metre run (in kilonewtons).

Stage 1. The section is in fact symmetrical so that the minor axis

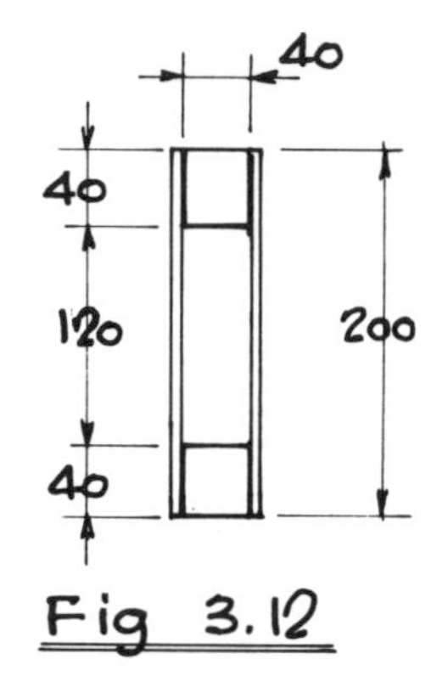

Fig 3.12

(Y–Y) is just the summation of 2 x $bd^3/12$ whereas the major axis (X–X) is the gross figure for $bd^3/12$ *minus* that of the void area. Thus:

$$I_{X-X} = BD^3/12 - bd^3/12$$

$$= \frac{40 \times 200^3 - 40 \times 120^3}{12}$$

$$= 10^6 \frac{(40 \times 2^3 - 40 \times 1.2^3)}{12}$$

$$= 10^6 \frac{(320 - 69.12)}{12} = 20.91 \times 10^6 \text{ mm}^4$$

$$I_{Y-Y} = \frac{2 \times 40 \times 40^3}{12}$$

$$= 10^6 \frac{(2 \times 40 \times 0.4^3)}{12} = 0.43 \times 10^6 \text{ mm}^4$$

Stage 2. To determine r_{X-X} and r_{Y-Y} values

$$r_{X-X} = \sqrt{\frac{I_{X-X}}{A}} = \sqrt{\frac{2091 \times 10^4}{2 \times 40 \times 40}}$$

$$= \sqrt{\frac{2091}{2 \times 0.4^2}} = 80.84 \text{ mm}$$

$$r_{Y-Y} = \sqrt{\frac{I_{Y-Y}}{A}} = \sqrt{\frac{43 \times 10^4}{2 \times 40^2}}$$

$$= \sqrt{\frac{43}{2 \times 0.4^2}} = 11.59 \text{ mm}$$

Stage 3. Theoretically, if the l/r value is to be equal for both axes, the effective height for the minor axis would need to be 11.59 ÷ 80.84 x 2400 = 344 mm, but this would be uneconomic. A stiffening stud spacing of 600 mm is the most frequent use of studding that might be envisaged. Thus the governing effective height is for the minor axis with an l/r value of 600 ÷ 11.59 = 51.8 (say 52). (*Note:* the reason for rounding up is that we will be using an imprecise material and interpolating on a table of approximations.)

Stage 3. Using an extract from CP112, table 15:

Slenderness ratio	k_{18} *value medium-term load*
40	1.13
50	1.08
60	1.00

we can interpolate so that 52 gives a reduction of 0.8 ÷ 5 = 0.016 from the k_{18} value of 1.08. Thus for l/r of 52 k_{18} = 1.08 - 0.016 = 1.064. Allowable compressive stress = 1.1 x 1.064 x 5.5 = 6.44 N/mm².

Stage 4

Safe load per metre = stress x $\frac{1000}{600} \times \frac{2 \times 40 \times 40}{10^3}$ kN/m

$$= \frac{6.44 \times 10^3 \times 3200}{600 \times 10^3} = 34.35 \text{ kN/m}$$

Answer: 34 kN/m run (rounded *down* for safety).

Should you wonder *why* an 'expanded' stud of this kind might be used, the answer is that it is a question of economics. Although, in practice, this section is no stronger than a solid section of 80 x 40 mm it does have certain advantages. If an external stud wall is required to house thermal insulation, the solid section can only take 80 mm thickness whereas the expanded section can accommodate 200 mm thickness, providing a much more effective resistance to heat loss through the fabric. Also, as mentioned earlier, no drilling is required for services such as gas, water and electricity. This example also shows the interdependence of mathematics, science and construction, since all three are considered.

Exercises

3.1 Determine the centroid for the figure shown in fig. 3.13 and indicate what its application to construction might be.

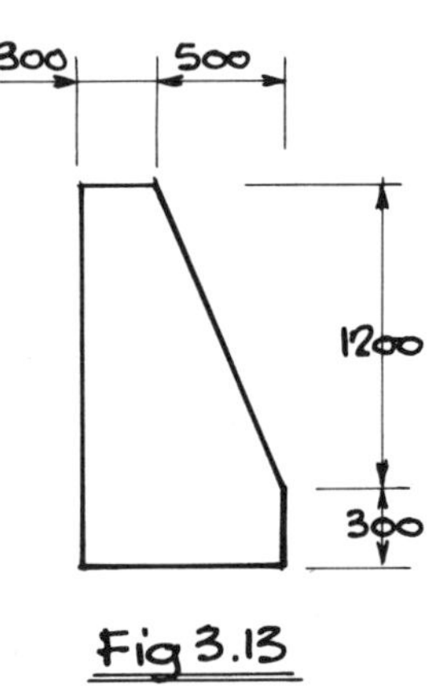

Fig 3.13

3.2 A cross-section through a concrete ribbed floor is shown in fig. 3.14. Determine the location of its neutral axis X–X.

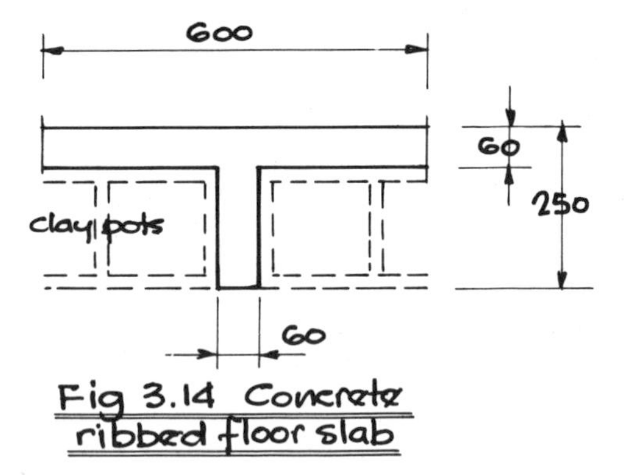

Fig 3.14 Concrete ribbed floor slab

3.3 A timber box beam of cross-section as shown in fig. 3.15 has solid ply webs. Determine the I_{X-X} and Z_{X-X} values for the section and, knowing that compressive stress is lower than tensile stress in allowable bending, explain how the misuse of this section could be dangerous.

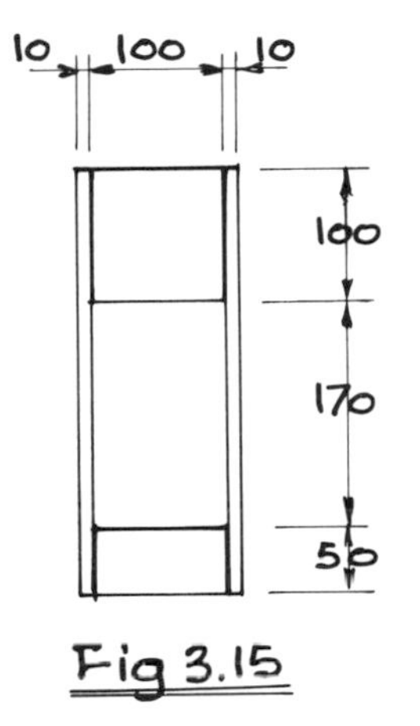

Fig 3.15

3.4 Figure 3.16 shows the cross-section of a pressed metal 'Z' purlin. Determine its I_{X-X} and Z_{X-X} values.

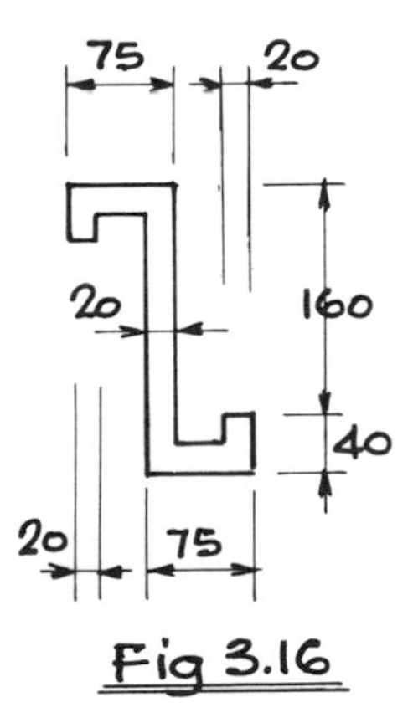

Fig 3.16

3.5 A precast concrete beam section as shown in fig. 3.17 is tapered to improve its resistance to compression in bending. Determine its I_{X-X} and Z_{X-X} values and compare them with a rectangular section of similar depth and area. (I for a triangle = $bd^3/36$.)

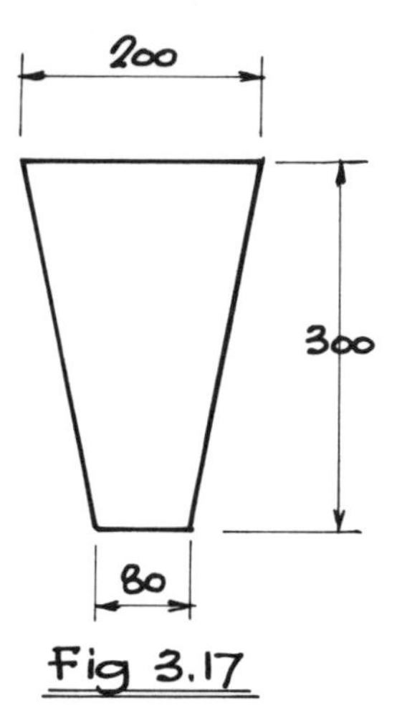

Fig 3.17

Assignment 3.1

A 'stressed skin' composite floor system of ply boarding and softwood joists is constructed to form a monolithic structural unit (i.e. one that acts as one piece of material). Given an E value of 10 000 N/mm^2 and a safe bending stress of 12 N/mm^2, determine the safe loading in kN/m^2 for floor spans of 4.0 m, 4.6 m and 5.2 m for the section shown in fig. 3.18. BM max. = WL/8 Deflection max. = $5WL^3/384EI$.

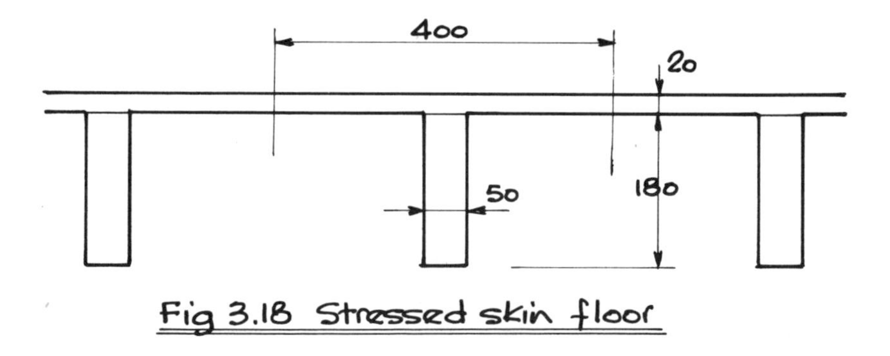

Fig 3.18 Stressed skin floor

4 Differential Calculus

In Volume 1, when we discussed graphs we showed how we could interpret information from them. When plotting a graph of a function in terms of x and y, we were able to establish the effect on x of changes of the value of y and vice versa. For example, if we know the gradient of a hill as 1 in 12, this can be expressed as y is 1 when x is 12 on a straight-line graph. Thus, if x is 36 m (the distance travelled), then y will equal 3 m (the height reached). Unfortunately, hills rarely climb at a uniform gradient but are made up of a series of gradients that form a curve. This means that the gradient at any point on the curve would be found by drawing a tangent to the curve *at that point*, which can be both tedious and inaccurate. However, if we know the equation of the curve, we can calculate its gradient at any point by using *differentiation*, which is a form of *calculus*.

Calculus is a classical form of mathematics and has, in many areas, been superseded by computing, which allows us to use *iterative* methods (or calculated guesswork) to determine answers by trial and error. However, calculus does still have its uses, particularly in its more simple form, and it is worth developing an understanding of its basic concepts.

BASIC DIFFERENTIATION

Returning to our graph with its variable gradient, in order to calculate the *rate of change* we must consider two points that are very close together. To do this we express the distances as δx and δy to indicate a *minute* increase in x and y. This is shown in fig. 4.1, which represents a graph of $y = x^2$. The minute increase is shown, in an exaggerated form, by a right-angled triangle which starts at point A on the curve and finishes at point B. The distances δx and δy are shown as the base and vertical height, respectively. Since

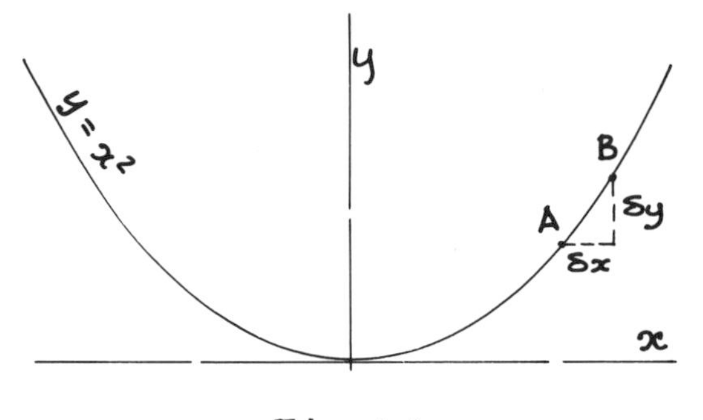

Fig 4.1

A is the point x, y (these being its ordinates), then B is the point $x + \delta x$, $y + \delta y$. The gradient is therefore $[(x + \delta x) - x]/[(y + \delta y) - y]$. Now, since $y = x^2$ (the equation of the graph), then $y + \delta y = (x + \delta x)^2$, since it follows the equation. When expanded this becomes $y + \delta y = x^2 + 2x.\delta x + \delta x^2$, and substituting x^2 for y, we obtain $x^2 + \delta y = x^2 + 2x.\delta x + \delta x^2$ so that $\delta y = 2x.\delta x + \delta x^2$. The gradient, therefore, being height ÷ distance, equals $\delta y/\delta x = 2x + \delta x$, and because the distances concerned are *extremely small*, we can ignore the contribution of δx so that $\delta y/\delta x = 2x$. If point B is ultimately made to coincide with point A ($\delta x \rightarrow 0$) the ratio of $\delta y/\delta x$ is given by the symbol dy/dx. This is known as the differential of y with respect of x. *Note*: it must be realised that δy and δx are *expressions* and the δ cannot be separated. Thus we cannot say $dy/dx = y/x$ since this is an *inadmissible statement.*

It is not necessary to develop the expansion for $\delta y/\delta x$ every time you wish to differentiate, because it follows a standard form so that when $y = ax^n$, $dy/dx = n.ax^{n-1}$. That is, the *original index power* becomes the *multiplier* and the new *index power* is the *original reduced by 1*. For example: $y = 3x^2$ when differentiated with respect to x becomes – $dy/dx = 2 \times 3x^{2-1} = 6x$ (since $x^1 = x$) and $y = 2x$ when differentiated with respect to x becomes $dy/dx = 1 \times 2x^{1-1} = 2x^0$ or 2 (since $x^0 = 1$). The standard form can also be applied to negative indices since $3/x$ is the same as saying $3x^{-1}$, so that if $y = 3/x$ is differentiated with respect to x, then $y = 3x^{-1}$ becomes $dy/dx = -1 \times 3x^{-1-1} = -3x^{-2}$ or $-3/x^2$.

It is interesting to see what happens when a constant is differentiated, i.e. a number with *no* function of x. We know that $x^0 = 1$, so 5 could be written as $5x^0$ if we wished. Thus $y = 5$ would be $y = 5x^0$ so that $dy/dx = 0 \times 5x^{0-1} = 0\ x^{-1}$. Since anything multiplied by 0 = 0, then the answer is $dy/dx = 0$.

DIFFERENTIATION OF A SUM

Let us look at what happens when an expression involving two or more terms, which are summated, is subjected to differentiation.

$$y = ax^2 + bx + c$$
$$y + \delta y = a(x + \delta x)^2 + b(x + \delta x) + c$$
$$= a(x^2 + 2x.\delta x + \delta x^2) + b(x + \delta x) + c$$
$$= ax^2 + 2ax\delta x + a\delta x^2 + bx + b\delta x + c$$

Subtracting y from each side: $\delta y = 2ax\delta x + a\delta x^2 + b\delta x$, so $\delta y/\delta x = 2ax + a\delta x + b$, which, since δx is *extremely small*, becomes $dy/dx = 2ax + b$. You can see that we could, in fact, have applied the standard form equally well to this expression since the constant (c) would have disappeared and the remainder would have become $2 \times ax^{-1} + 1 \times bx^0$, thus $dy/dx = 2ax + b$.

SECOND-ORDER AND HIGHER-ORDER DERIVATIVES

If $y = f(x)$, then y may be differentiated with respect to x, and since dy/dx gives the rate of increase of y with respect to x, then the procedure of differentiation may be carried out successively, i.e. as y is a function of x, so dy/dx is also a function of x. The derivative of dy/dx is $\dfrac{d(dy/dx)}{dx} = \dfrac{d^2y}{dx^2}$ which is called the *second derivative of* y with respect to x. This can be repeated successively through the higher orders so that if $y = f(x)$ dy/dx may be written as $f'(x)$, d^2y/dx^2 as $f''(x)$ and so on.

You may wonder where all this is leading, so it is well worth returning to our graphs. We talked of $\delta y/\delta x$ as being a means of expressing a gradient, but in some cases we are just as interested to know the conditions when no gradient occurs, i.e. at a change of direction on the graph. Figure 4.2 shows conditions of rate of change where in the first case y increases as x increases, so that the tangent slopes *upwards* from left to right. In the second case, y decreases as x increases, so that the tangent slopes *downwards* from left to right, whereas in the third case, midway between cases 1 and 2, the curve changes direction, so that the tangent is *horizontal*. This would be equally true if the curve had 'bottomed out', i.e. between cases 2 and 1. These are known as *rates of change*, and if

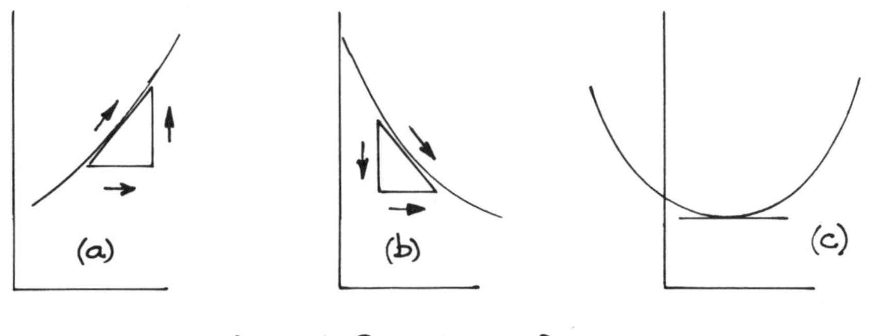

Fig 4.2 rates of change

we apply this principal to a cubic graph, as shown in fig. 4.3, in using the second derivatives they will tell us the same things about $\delta y/\delta x$ that the first derivatives told us about y.

The points on the curve when $dy/dx = 0$ are called *stationary points*, and also represent the maximum and minimum values of y provided that the sign of the differential coefficient changes from negative to positive (or positive to negative). It will, therefore, be seen that when y is a *maximum*, dy/dx, must equal zero with d^2y/dx^2 *negative*, and that when y is a *minimum*, dy/dx must equal zero with d^2y/dx^2 *positive*.

Example 4.1
Find maximum and minimum values for y when $y = 3x^3 - 8x^2 - 4x + 3$.

$$dy/dx = 9x^2 - 16x - 4$$

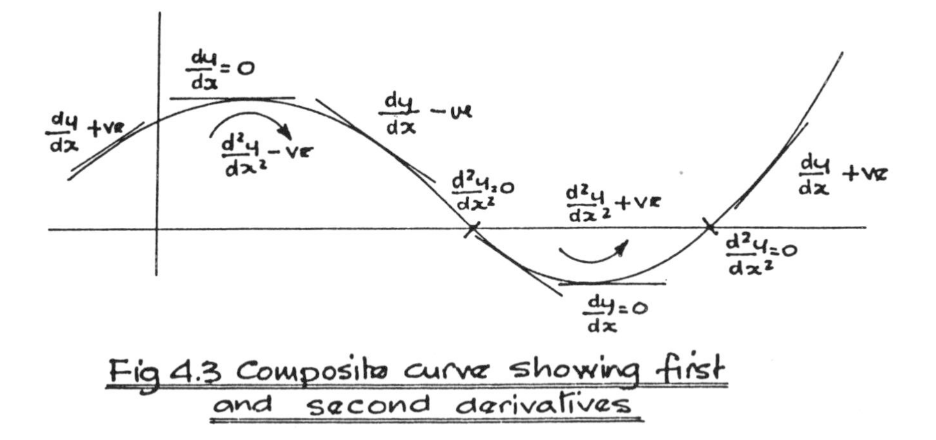

Fig 4.3 Composite curve showing first and second derivatives

so that if $dy/dx = 0$, then $0 = 9x^2 - 16x - 4$, which factorises to $(9x + 2)(x - 2) = 0$

$$x = -0.2\dot{2} \text{ or } +2.0$$

Taking a second differentiation, $d^2y/dx^2 = 18x - 16$, which, when substituting for $x = -0.2\dot{2}$, produces $-18 \times 0.2\dot{2} - 16 = -4 - 16 = -20$ (negative so y is maximum) or, when substituting for $x = +2.0$, produces $18 \times 2 - 16 = 36 - 16 = +20$ (positive so y is minimum).

We can now substitute back into the original cubic equation $y = 3x^3 - 8x^2 - 4x + 3$ for both values of x. Maximum ($x = -0.2\dot{2}$):

$$y = 3(-0.2\dot{2})^3 - 8(-0.2\dot{2})^2 - 4(-0.2\dot{2}) + 3 = -0.033 - 0.395 + 0.889 + 3 = 3.461$$

and for minimum ($x = +2.0$):

$$y = 3(2.0)^3 - 8(2.0)^2 - 4(2.0) + 3 = 24 - 32 - 8 + 3 = -13$$

It is, of course, not critical that we use cubic equations in relation to maxima and minima, which has its uses in terms of costing and estimating.

Example 4.2

The cost of an electric cable per kilometre is given by the formula: $C = \frac{180\,000}{x} + 20x$, where C is its cost in £ and x is its cross-sectional area in square millimetres. Find the cross-section for which the cost is least, and the least cost per kilometre.

$$C = \frac{180\,000}{x} + 20x$$

$$dc/dx = -\frac{180\,000}{x^2} + 20$$

For a minimum (or maximum) dc/dx must be zero, therefore:

$0 = -180\,000/x^2 + 20$, which, transposing, gives
$180\,000 = 20x^2$

$x = \pm\ 94.87\ \text{mm}^2$

(core diameter = 11 mm)

Now substitute back into original to find C.

$$C = \frac{180\,000}{94.87} + 20 \times 94.87$$

$$= 1897.33 + 1897.4$$

$$= 3794.73, \quad \text{say £3795}$$

In practice it would probably be priced at £3 800 per kilometre.

Example 4.3

This is a more traditional, if slightly impractical, problem relating surface area to volume. A glass-fibre cold water storage tank has base dimensions $x \times 1.2x$ and a height of y (all dimensions in metres). Determine the maximum volume (to the brim) for a surface area 3.0 m².

First we establish the surface area in terms of x and y which gives $2 \times 1.2xy + 2 \times xy + 1.2x^2$ (open top).

$$3.0 = 4.4xy + 1.2x^2$$

Transposing to find y:

$$3 - 1.2x = 4.4xy$$

$$\frac{3 - 1.2x^2}{4.4x} = y$$

Equation (i) is $\quad y = \dfrac{0.682}{x} - 0.273x$

Equation (ii) is $\quad \text{Volume} = 1.2x^2 y$

Substitute for y in equation (ii):

$$\text{Volume} = 1.2x^2 \left(\frac{0.682}{x} - 0.273x\right)$$

$= 0.818x - 0.328x^3$

$dv/dx = 0.818 - 0.984x^2$

and since, for the max volume $dv/dx = 0$

$$x = \sqrt{\frac{0.818}{0.984}}$$

$= 0.911$ m

Substituting back into equation (i):

$$y = \frac{0.682}{0.911} - 0.273 \times 0.911$$

$= 0.749 - 0.249$

$= 0.50$ m

Volume $= 1.2 \times 0.911^2 \times 0.5$

$= 0.498\ m^3$

In practice the final dimensions would be rounded to 900 x 1100 x 500 mm, giving a surface area of $2 \times 0.9 \times 0.5 + 2 \times 1.1 \times 0.5 + 0.9 \times 1.1 = 2.99\ m^2$, and a volume of $0.9 \times 1.1 \times 0.5 = 0.495\ m^3$.

There are several other techniques that we may need to use in differentiation. These cover trigonometric, logarithmic and exponential functions. However, these are seldom used in the solution of construction problems and we shall concern ourselves only with those used for algebraic functions. (*Note:* should you particularly wish to develop your own knowledge of calculus further, there are many books that cover the subject in depth.)

FUNCTION OF A FUNCTION

Not all equations follow the straightforward format that we have seen in those concerned primarily with sum and difference statements. Let us examine how we approach an equation such as:

$y = (2x^3 + 4x^2 - 5x - 3)^2$

We could, of course, multiply out the expression to clear the brackets, but this would be time-consuming and prone to error. Instead we will consider it in two stages. Let $p = 2x^3 + 4x^2 - 5x - 3$, so that $y = p^2$.

Then $dy/dp = 2p$

$= 2(2x^3 + 4x^2 - 5x - 3)$

but $dp/dx = 6x^2 + 8x - 5$

so that $dy/dp \times dp/dx = dy/dx$

thus $dy/dx = 2(2x^3 + 4x^2 - 5x - 3)(6x^2 + 8x - 5)$

What we have done, therefore, is to split the problem into two elements, treating each as a function, hence the use of the term 'function of a function'. By differentiating p with respect to y, and then x with respect to p, we have eased the problem and only have to multiply both differentials together, i.e.

Stage 1: dy/dp

Stage 2: dp/dx

Stage 3: $dy/dp \times dp/dx = \dfrac{dy}{\cancel{dp}} \times \dfrac{\cancel{dp}}{dx} = \dfrac{dy}{dx}$

Example 4.4

Differentiate with respect to x;

$$y = \sqrt{x^2 - 1}$$

Let $p = x^2 - 1$ so that $y = \sqrt{p}$ or $y = p^{½}$.

$$dy/dp = ½p^{-½} \text{ or } 0.5/\sqrt{p}$$

$$\text{Thus } dy/dp = \frac{0.5}{\sqrt{x^2 - 1}}$$

$$dp/dx = 2x$$

$$\text{Therefore } dy/dx = \frac{2x \times 0.5}{\sqrt{x^2 - 1}}$$

$$= \frac{x}{\sqrt{x^2 - 1}}$$

PRODUCT RULE

This applies where the expression is of a form involving the product of two functions, i.e.

$$y = (2x^3 + 3x^2)(x^4 + 1)$$

Again, as before, we could multiply out to clear the brackets, but if we consider each bracket as a more simple unit $y = u \times v$, where $u = 2x^3 + 3x^2$, and $v = x^4 + 1$, we can use the expression:

$$\frac{dy}{dx} = \frac{u.dv}{dx} + \frac{v.du}{dx}$$

Thus, to solve the problem, we move in easy stages:

Stage 1: $\frac{dv}{dx} = 4x^3$

Stage 2: $u.\frac{dv}{dx} = (2x^3 + 3x^2)(4x^3)$

Stage 3: $\frac{du}{dx} = 6x^2 + 6x$

Stage 4: $v.\frac{du}{dx} = (x^4 + 1)(6x^2 + 6x)$

Thus, summating $u.\frac{dv}{dx} + v.\frac{du}{dx}$:

$$\frac{dy}{dx} = 4x^3(2x^3 + 3x^2) + (x^4 + 1)(6x^2 + 6x)$$

We may simplify this to $8x^6 + 12x^5 + 6x^6 + 6x^5 + 6x^2 + 6x$.

Thus $\frac{dy}{dx} = 14x^6 + 18x^5 + 6x^2 + 6x$

As you can see, it isn't always quicker to use such techniques since the original equation was easy to multiply, producing $y =$

$2x^7 + 3x^6 + 2x^3 + 3x^2$, giving $dy/dx = 14x^6 + 18x^5 + 6x^2 + 6x$. It does, however, help to prove that the product rule does work!

Example 4.5
Differentiate with respect to x:

$$y = (2x^3 - 3)(4x^2 + 5)$$

$$u = 2x^3 - 3$$

$$v = 4x^2 + 5$$

$$\frac{dv}{dx} = 8x$$

$$u.\frac{dv}{dx} = 8x(2x^3 - 3) = 16x^4 - 24x$$

$$\frac{du}{dx} = 6x^2$$

$$v.\frac{du}{dx} = 6x^2(4x^2 + 5) = 24x^4 + 30x^3$$

$$\frac{du}{dx} = 16x^4 - 24x + 24x^4 + 30x^3$$

$$= 40x^4 + 30x^3 - 24x$$

QUOTIENT RULE

This applies where the expression is of a form involving the quotient of two functions, i.e.

$$y = \frac{2x^3 + 3x^2}{x^4 + 1}$$

It is less easy to divide out in this instance so it is useful to know the rule where

$$y = \frac{u}{v}$$

and $\dfrac{dy}{dx} = \dfrac{v.\ du/dx - u.\ dv/dx}{v^2}$

We already know from the product rule that if $u = 2x^3 + 3x^2$, and $v = x^4 + 1$, then $v.\ \dfrac{du}{dx} = (x^4 + 1)(6x^2 + 6x)$, which simplifies to $6x^6 + 6x^5 + 6x^2 + 6x$, and $u.\ \dfrac{dv}{dx} = 4x^3(2x^3 + 3x^2)$, which simplifies to $8x^6 + 12x^5$, and we can establish v^2 (where $v = x^4 + 1$) as $v^2 = x^8 + 2x^4 + 1$.

This gives $\dfrac{v.\ du/dx - u.\ dv/dx}{v^2}$

$$= \frac{6x^6 + 6x^5 + 6x^2 + 6x - 8x^6 - 12x^5}{x^8 + 2x^4 + 1}$$

$$= \frac{-2x^6 - 6x^5 + 6x^2 + 6x}{x^8 + 2x^4 + 1}$$

or $\dfrac{6x + 6x^2 - 6x^5 - 2x^6}{1 + 2x^4 + x^8}$

Note: it does not matter in which order the powers of x are arranged, so long as they are consecutive. Since it is preferable *not* to start an expression with a minus sign, the order was reversed.

Example 4.6
Differentiate with respect to x:

$$y = \frac{3x - 4}{2x^2 + 1}$$

Stage 1: $u = 3x - 4$
Stage 2: $du/dx = 3$
Stage 3: $v = 2x^2 + 1$
Stage 4: $dv/dx = 4x$
Stage 5: $v^2 = 4x^4 + 4x^2 + 1$

Therefore $v.\,du/dx = (2x^2 + 1)(3) = 6x^2 + 3$
$u.\,dv/dx = (3x - 4)(4x) = 12x^2 - 16x$
$v.\,du/dx - u.\,dv/dx = 6x^2 + 3 - 12x^2 + 16x$
$= 3 + 16x - 6x^2$

$$dy/dx = \frac{3 + 16x - 6x^2}{1 + 4x^2 + 4x^4}$$

All the expressions with which we have dealt so far have been *explicit functions*, where it has been possible to isolate y in terms of x. This is not always possible, although it is sometimes a question of transposing an equation to make it explicit, and where y occurs in a variety of index powers, the expression is known as an *implicit function*, e.g.

$$x^2 + y^2 + 3y - 7 = 0$$

IMPLICIT FUNCTIONS

To differentiate an implicit function we transpose the terms involving y to one side of the equation and the remainder, including constants, to the opposite side. We then treat the x and y components quite separately and, depending on the complexity of those involving y, consider these either individually or as an expression adopting the 'function of a function' technique. The equation can then be resolved to find the dy/dx value. Let us look at the equation:

$$x^2 + y^2 + 3y - 7 = 0$$

and adopt the alternative approaches, as outlined above. For both methods we should first transpose the equation to read:

$$y^2 + 3y = 7 - x^2$$

Method 1: let $y^2 = p$, and $3y = q$

$$dp/dx = dp/dy \times dy/dx \text{ and } dq/dx = dq/dy \times dy/dx$$

Thus $dp/dy = 2y$ so $dp/dx = 2y.dy/dx$

$dq/dy = 3$ so $dq/dx = 3.dy/dx$

and dy/dx for $7 - x^2 = -2x$

The equation becomes

$$2y.dy/dx + 3.dy/dx = -2x$$

Therefore $$dy/dx = \frac{-2x}{2y+3}$$

Method 2: Let $y^2 + 3y = p$

$$dp/dx = dp/dy \times dy/dx$$

$$= (2y+3)\,dy/dx$$

$$dy/dx \text{ for } 7 - x^2 = -2x$$

Therefore $$(2y+3)\,dy/dx = -2x$$

$$dy/dx = \frac{-2x}{2y+3}$$

Note: when an expression has been transposed, it may be necessary to employ one of the rules covered earlier, e.g. product or quotient for terms involving x.

Example 4.7
Differentiate with respect to x:

$$x^2 + xy + 2 = 0$$

Transpose to read: $xy = -2 - x^2$

This can now be written as:

$$y = \frac{-(2+x^2)}{x}$$

and differentiated *explicitly*, thus:

let $u = -(2+x^2)$

$$v = x$$

using quotient rule $\dfrac{dy}{dx} = \dfrac{v.du/dx - u.dv/dx}{v^2}$

$$\frac{du}{dx} = -2x$$

Therefore $v.\,du/dx = -2x^2$

$$\frac{dv}{dx} = 1$$

Therefore $u.\,dv/dx = -(2 + x^2)$

$$v.\,du/dx - u.\,dv/dx = -2x^2 + 2 + x^2$$

$$= 2 - x^2$$

Therefore $dy/dx = (2 - x^2)/x^2$

Alternatively, should we have retained xy as an expression, we would have to differientiate *implicitly*, thus:

$$xy = -2 - x^2$$

Let $u = x$ and $v = y$

$$du/dx = 1 \quad \text{so} \quad v.\,du/dx = y$$

$$dv/dx = 1.dy/dx \quad \text{so} \quad u.dv/dx = x.dy/dx$$

The equation thus reads:

$$x.dy/dx + y = -2x$$

$$x.\,dy/dx = -2x - y$$

Therefore $dy/dx = (-2x - y)/x$

which is still *implicit* and, though different in its appearance from the first answer, should be similar in value. *Note:* we can, in fact, check this to be the case since if both expressions were brought to the same denominator, $1/x^2$, we would assume that

$$\frac{2 - x^2}{x^2} = - \frac{2x^2 - xy}{x^2}$$

or $2 - x^2 = - 2x^2 - xy$

which, when transposed, equals $xy = - 2 - x^2$, which was our original equation. Therefore both answers are equally correct.

Exercises
Differentiate with respect to x:

4.1 $y = 4x^2 - 3x + 2$
4.2 $y = \sqrt{x}$
4.3 $4x^2 + y = 3$
4.4 $y = 1 + 1/x$
4.5 $y = (x - 3)(x + 2)$ by first clearing brackets
4.6 $y = (x^2 + 3)^2$ as a function of a function
4.7 $y = (x - 3)(x + 2)$ using the product formula
4.8 $y = (x^2 + 3)^2$ using the product formula
4.9 $y = (2x - 1)/(x^2 + 2)$
4.10 $y^2 + 4y - 3x + 2 = 0$

5 Integral Calculus

Let us start with the question 'What is integration?' In its most literal sense integration is the collection of all the parts into a whole, as in semi-graphical integration when we summate areas. In its mathematical sense it is the determination of a function from its differential coefficient. So in simple terms we can think of it as the reverse of differentiation.

Thus, if we know, from chapter 4, that when $y = x^4/4$ $dy/dx = 4x^3/4 = x^3$, we can equally well say that $dy = x^3 dx$. The integral of $x^3\,dx$ will therefore be $x^4/4$. We use a Greek letter to identify the term 'integral', which we described above as the summating of parts into a whole. We know from structural mechanics that the capital sigma (Σ) is used for manual summation, so we use the lower case ($\int$) in the form of an integral sign for integration by calculus, i.e.

$$\int x^3\,dx = x^4/4$$

You will notice that, whereas with differentiation we *multiplied* by the index power and then *reduced* its value by 1, with integration we *raise* the index power by 1 and then *divide* by the *new* index power. Thus:

$$\int ax^n\,dx = \frac{a}{(n+1)} . x^{(n+1)}$$

Note: this is a general rule *except* for $\int x^{-1}\,dx$, which equals $\log_e x$

BASIC INTEGRATION

We know from chapter 4, and from the introduction above, that differentiating $y = x^2$ gives $dy/dx = 2x$ and that $\int 2x.dx = 2x^{(1+1)}/2$

$= x^2$. However, we also know that differentiating $y = x^2 + 2$ or $x^2 + 15$ would also give $dy/dx = 2x$, since the constant will disappear in the differentiation. In practice, therefore, we must allow for the possibility of this *constant* by adding a symbol (k) to the integrated form. Thus:

$$\int ax^n\, dx = \frac{a}{(n+1)} . x^{(n+1)} + k$$

You will see from the above that the *coefficient* of $x(a)$ is unaffected by the integration and so the expression may be written alternatively as $a\int x^n \delta x$, e.g.

$$\int 5x^3\, dx = 5\int x^3\, dx = \frac{5}{4} x^{(3+1)} + k = \frac{5}{4} x^4 + k$$

In the same way as for differentiation we can 'prepare' an expression for integration by changing, for example, $1/x^2$ to x^{-2}. This means that $\int 1/x^2 . dx$ can be written as

$$\int x^{-2} . dx = \frac{x}{-1}^{(-2+1)} + k = - 1/x + k$$

INTEGRATION OF A SUM

The rules for sums and differences that we applied to differentiation apply equally to integration whereby we consider each term separately so that

$$\int (ax^2 + bx + c)dx = \frac{a.x^3}{3} + \frac{bx^2}{2} + cx + k$$

Example 5.1
Integrate the expression $3x^2 - 4x + 2$.

$$\int (3x^2 - 4x + 2)dx = \frac{3x^{(2+1)}}{3} - \frac{4x^{(1+1)}}{2} + \frac{2x^{(0+1)}}{1} + k$$

$$= x^3 - 2x^2 + 2x + k$$

Example 5.2
Integrate the expression $\frac{5}{x^6} + 3x^4 - \frac{6}{x^2} + 2x$

$$\int(5x^{-6} + 3x^4 - 6x^{-2} + 2x)dx$$

$$= \frac{5x^{(-6+1)}}{-5} + \frac{3x^{(4+1)}}{5} - \frac{6x^{(-2+1)}}{-1} + \frac{2x^{(1+1)}}{2} + k$$

$$= -x^{-5} + \frac{3}{5}x^5 + 6x^{-1} + x^2 + k$$

which we express, in index order, as

$$\frac{3x^5}{5} + x^2 + \frac{6}{x} - \frac{1}{x^5} + k$$

FINDING THE CONSTANT OF INTEGRATION

In the examples above we have considered the constant (k) as an arbitrary figure. When we *cannot* establish a true value for k, the expression is known as an *indefinite* integral because we do not know the *overall* value of the resultant expression. If, however, we know of a condition that provides corresponding values of x and y, we *can* establish a true value for k and, in these circumstances, the expression is known as a definite integral because we can work out the overall value of the resultant expression.

Example 5.3
The gradient of a curve which passes through the point ($x = 1$, $y = 2$) is given by the expression x^2. We know that the equation for the curve can be found by integrating its gradient, thus:

$$y = \int x^2 . dx = \frac{x^3}{3} + k$$

By substituting values for x and y we produce $2 = 1^3/3 + k = 1/3 + k$, which, when transposed, gives $2 - \frac{1}{3} = k$. Therefore $k = 1\frac{2}{3}$ or $\frac{5}{3}$. The equation for the curve, therefore, is

$$y = \frac{x^3}{3} + \frac{5}{3} \text{ or } \tfrac{1}{3}(x^3 + 5)$$

This will be of use when we apply integration to structural problems. Alternatively, we can eliminate the constant when we integrate an expression between limits, i.e. if we limit the possible values of x between the range of a and b, thus:

$$\int_b^a x^n \, dx,$$

we produce:

$$\left[\frac{x^{(n+1)}}{(n+1)} + k\right]_b^a$$

for which we can then substitute the limiting values to produce an answer. This is also a *definite* integral because we can find an overall value.

Example 3.4: Integrating between limits

$$\int_1^3 (2x^3 + x^2 - x + 3)dx$$

$$= \left[\frac{2x^{(3+1)}}{4} + \frac{x^{(2+1)}}{3} - \frac{x^{(1+1)}}{2} + \frac{3x^{(0+1)}}{1} + k\right]_1^3$$

$$= \left[\frac{x^4}{2} + \frac{x^3}{3} - \frac{x^2}{2} + 3x + k\right]_1^3$$

We now substitute values for $x = 3$ and subtract a second expression for $x = 1$. That is:

$$\left(\frac{x^4}{2} + \frac{x^3}{3} - \frac{x^2}{2} + 3x + k\right)x = 3 - \left(\frac{x^4}{2} + \frac{x^3}{3} - \frac{x^2}{2} + 3x + k\right)x = 1$$

$$= \left(\frac{81}{2} + \frac{27}{3} - \frac{9}{2} + 9 + k\right) - \left(\frac{1}{2} + \frac{1}{3} - \frac{1}{2} + 3 + k\right)$$

$$= (54 + k) - (3\tfrac{1}{3} + k)$$

The k values cancel out, being unaffected by the limiting values for x, to give $50\tfrac{1}{3}$. This works equally well if one of the values for x is negative.

Example 5.5

$$\int_{-2}^{2} (3x^2 + 2x - 4)dx$$

$$= \left[\frac{3x^{(2+1)}}{3} + \frac{2x^{(1+1)}}{2} - \frac{4x^{(0+1)}}{1} + k\right]_{-2}^{2}$$

$= \left[x^3 + x^2 - 4x + k\right]_{-2}^{2}$, and since we can now ignore k, which is eliminated, this produces

$$(8 + 4 - 8) - (-8 + 4 + 8)$$

$$= 4 - 4 = 0$$

SECOND-ORDER AND HIGHER-ORDER INTEGRATION

Just as we can differentiate successively, we can also integrate successively, but in doing so we must be able either to determine or eliminate the constant at each stage. We represent this process by $\int x.dx$, $\int\int x.dx$, $\int\int\int x.dx$ or $\int' x.dx$, $\int'' x.dx$, $\int''' x.dx$, though the second format can be confusing if limits are used.

The most common application of successive integration is in the analysis of beams. In *Construction Science* Volume 2 we looked at deflection of beams as a continuous process of examining loading, shear, bending, slope and deflection, and we can relate this sequence in mathematical terms to:

Shear = Integral of loading ($\int W$)
Bending = Integral of shear ($\int \text{Sh} = \int\int W$)
Slope = Integral of bending ($\int \text{B} = \int\int \text{Sh} = \int\int\int W$)
Deflection = Integral of slope ($\int \text{Sl} = \int\int \text{B} = \int\int\int \text{Sh} = \int\int\int\int W$)

Let us apply this to two of the more basic beam mechanisms, the central point load and the uniformly distributed load, both with simple supports at each end.

Example 5.6

Figure 5.1 shows the behaviour of a point-loaded beam system in shear, bending, slope and deflection. Knowing that the curvature

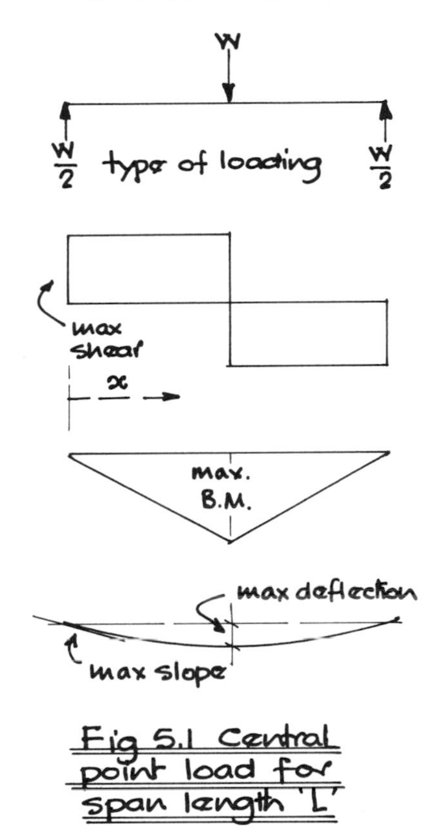

Fig 5.1 Central point load for span length 'L'

of the beam under load is $W/2 = EId^3 y/dx^3$, determine maximum values for bending slope and deflection using successive integration.

We can see from the shearing force diagram that the maximum force is consistent as $W/2$ across the entire span except directly under the load itself. Because we read from left to right, we should express the shearing force as $-W/2$ at the left-hand support since it is caused theoretically by the reaction. We need integrate only *three* times to find its deflection and can progress directly to bending.

Stage 1

$$\text{Bending} = \int \text{Shear} = \int -W/2.dx \;\; = -\; Wx/2 + k$$

Bending is zero at the support when $x = 0$, so

$$0 = -\; W \times 0/2 + k \quad k = 0$$

Therefore bending moment = $Wx/2$ maximum at centre, when $x = L/2$

$$= -WL/2 \times 2 = -WL/4$$

Stage 2

$$\text{Slope} = \int \text{Bending} = \int -\frac{Wx \, . dx}{2EI} = -\frac{Wx^2}{4EI} + k$$

(Note: we introduce the division EI at this point because of the rotation of the beam.)

Slope is zero directly under the point load when $x = L/2$, so

$$0 = -\frac{WL^2}{16EI} + k \qquad k = \frac{WL^2}{16EI}$$

$$\text{Slope} = -\frac{Wx^2}{4EI} + \frac{WL^2}{16EI} = \frac{W}{4EI}\left(-x^2 + \frac{L^2}{4}\right)$$

Maximum slope is at the support when $x = 0$, so maximum slope = $WL^2/16EI$.

Stage 3

$$\text{Deflection} = \int \text{Slope} = \frac{W}{4EI}\int\left(-x^2 + \frac{L^2}{4}\right) dx$$

$$= \frac{W}{4EI}\left(-\frac{x^3}{3} + \frac{L^2x}{4} + k\right)$$

Deflection is zero when x is zero, so $k = 0$.

$$\text{Deflection} = \frac{W}{4EI}\left(-\frac{x^3}{3} + \frac{L^2x}{4}\right)$$

Maximum deflection occurs at mid-span when $x = L/2$, so maximum deflection $= \frac{W}{4EI}\left(-\frac{L^3}{24} + \frac{L^3}{8}\right) = \frac{WL^3}{4EI}\left(\frac{-1+3}{24}\right)$

$$= \frac{WL^3}{48EI}$$

This corresponds to the value given in *Construction Science* Volume 2 (chapter 4).

Example 5.7

Figure 5.2 shows the behaviour of a uniformly loaded beam of load rate w kN/m. Knowing the curvature of the beam is $w = EId^4\, y/dx^4$, determine the maximum values for bending, slope and deflection.

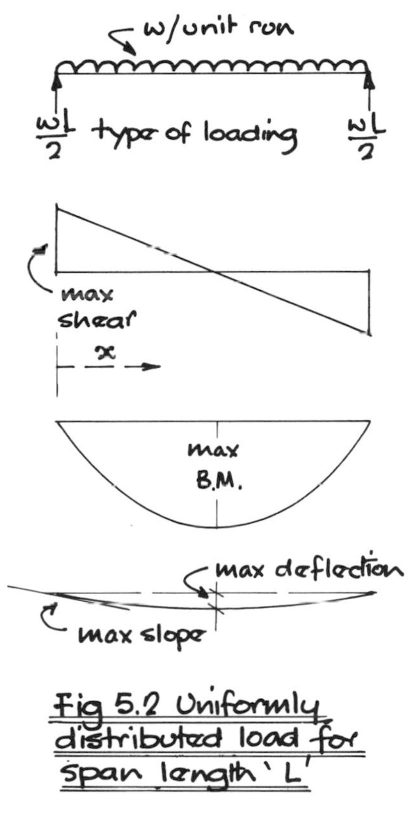

Fig 5.2 Uniformly distributed load for span length 'L'

Stage 1

$$\text{Shear} = \int \text{Load} = \int w\, dx = wx + k$$

We can see from the shearing force diagram that maximum shear occurs at the supports when $x = 0$ or L. Reading from the left, maximum $= -\ wL/2$, thus $-\ wL/2 = wx + k$ when $x = 0$ so $k = -\ wL/2$.

$$\text{Shear} = wx - \ wL/2 \text{ or } w(x - L/2)$$

Stage 2

Bending = $\int$Shear = $\int w(x - L/2)dx = w(x^2/2 - Lx/2 + k)$ but bending is zero when x is zero, so $k = 0$. Bending = $w(x^2/2 - Lx/2)$; maximum occurs at mid-span when $x = L/2$ so maximum bending moment = $w(L^2/8 - L^2/4) = - wL^2/8$.

Stage 3

$$\text{Slope} = \int \text{Bending} = \int \frac{w}{EI}\left(x^2/2 - Lx/2\right) dx$$

$$= w\left(\frac{x^3}{6} - \frac{Lx^2}{4} + k\right)$$

Slope is zero at mid-span, when $x = L/2$, so

$$0 = \frac{w}{EI}\left(\frac{L^3}{48} - \frac{L^3}{16} + k\right)$$

$$k = \frac{L^3}{16} - \frac{L^3}{48} = \frac{L^3}{24}$$

$$\text{Slope} = \frac{w}{EI}\left(\frac{x^3}{6} - \frac{Lx^2}{4} + \frac{L^3}{24}\right)$$

Maximum occurs at support when $x = 0$, so maximum slope = $wL^3/24EI$.

Stage 4

$$\text{Deflection} = \int \text{Slope} = \frac{w}{EI}\int\left(\frac{x^3}{6} - \frac{Lx^2}{4} + \frac{L^3}{24}\right) dx$$

$$= \frac{w}{EI}\left(\frac{x^4}{24} - \frac{Lx^3}{12} + \frac{L^3 x}{24} + k\right)$$

Deflection is zero at the support when $x = 0$, so $k = 0$

$$\text{Deflection} = \frac{w}{24EI}\left(x^4 - 2Lx^3 + L^3 x\right)$$

Maximum deflection occurs at mid-span when $x = L/2$, so

$$\text{maximum} = \frac{w}{24EI}\left(\frac{L^4}{16} - \frac{2L^4}{8} + \frac{L^4}{2}\right)$$

$$= \frac{wL^4}{24EI}\left(\frac{1-4+8}{16}\right) = \frac{5wL^4}{24 \times 16 \times EI}$$

$$= \frac{5wL^4}{384EI} \text{ or, since } W = wL, \ \frac{5WL^3}{384EI}$$

FUNCTION OF A FUNCTION

If we wish to integrate $(ax + b)^n$, we must remember the way we substituted a more simple function for the bracketed expression when we used differentiation, i.e. let $p = ax + b$ so that $y = p^n$.

Knowing that integration is the reverse of differentiation let us first differentiate $(ax + b)^{n+1}$ with respect to x.

$$p = ax + b \qquad y = p^{n+1}$$

$$dy/dp = (n + 1)p^n = (n + 1)(ax + b)^n$$

$$dp/dx = a \text{ so } dp/dx \times dy/dp = dy/dx = a(n + 1)(ax + b)^n$$

Now, if we integrate the result, we will return to the original, i.e.

$$\int a(n + 1)(ax + b)^n = (ax + b)^{n+1}$$

so that by transposing $a(n + 1)$ we obtain

$$\int (ax + b)^n = \frac{(ax + b)^{n+1}}{a(n + 1)}$$

The general expression, therefore, allowing for the constant, is:

$$\int (ax + b)^n = \frac{(ax + b)^{n+1}}{a(n + 1)} + k$$

Applying this, then, $\int (2x + 3)^3 = (2x + 3)^4/8 + k$

AREAS UNDER CURVES

Figure 5.3 shows a section of a graph whose curve shows y as a function of x. The limits that we are considering are A to B. The strip $PQRS$ represents a very narrow strip of breadth δx. Thus if $CR = x$, $CS = x + \delta x$, and if $PR = y$ then $QS = y + \delta y$.

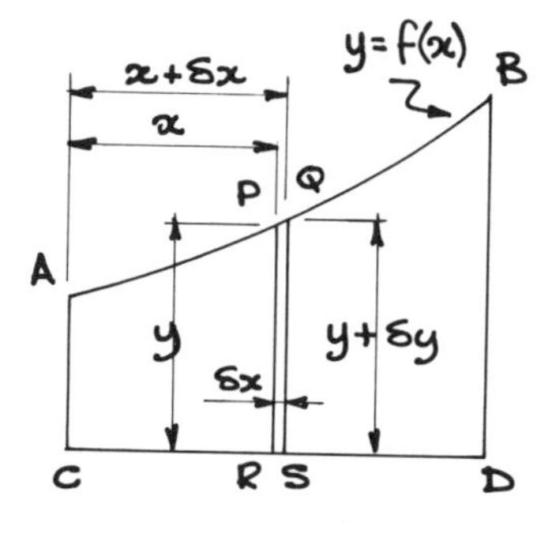

Fig 5.3

At the start of the chapter we said that integration can mean the summation of the parts to form the whole so that if we were to summate all the narrow strips of breadth δx between C and D, we could find the area under the curve (this was mentioned when we discussed the trapezoidal rule in chapter 2). Since all strips are of equal width (δx), then the variable is the y-ordinate so the area $= \int y.\delta x$, and since we know the limits as A and B, then the area shown can be determined through $\int_A^B y.dx$.

Example 5.8

Figure 5.4 shows a parabolic curve $y = x^2 + 2x + 1$ between the limits of -1 and 2. Determine the area under the curve and check the answer against the formula shown in Volume 1, chapter 8.

$$\text{Area} = \int_{-1}^{2} x^2 + 2x + 1 = \left[\frac{x^3}{3} + x^2 + x\right]_{-1}^{2}$$

$$= \left[\frac{x^3}{3} + x^2 + x\right]^{2} - \left[\frac{x^3}{3} + x^2 + x\right]^{-1}$$

$$= \left[\frac{8}{3} + 4 + 2\right] - \left[-\frac{1}{3} + 1 - 1\right]$$

$$= (2\tfrac{2}{3} + 6) - (-\tfrac{1}{3}) = 8\tfrac{2}{3} + \tfrac{1}{3} = 9 \text{ units}^2$$

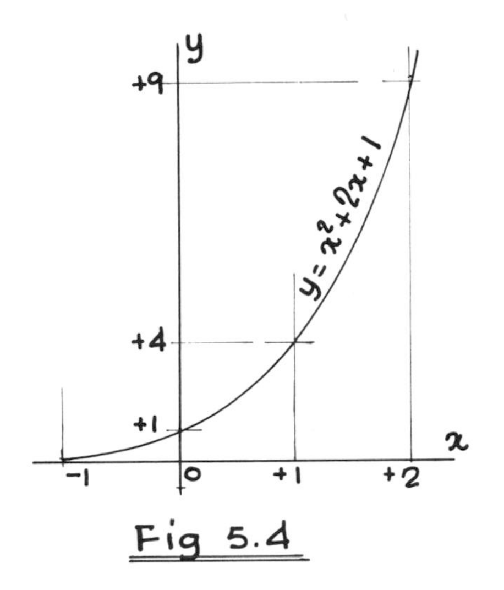

Fig 5.4

From Volume 1, area under a minor parabola = (height x base)/3, where height = 9 and base = 3 (distance - 1 to 2). Therefore area = 9 x 3 ÷ 3 = 9 units².

Example 5.9
Figure 5.5 shows a circle of radius R and a strip of breadth δr at a radius r from the centre. Determine its area.

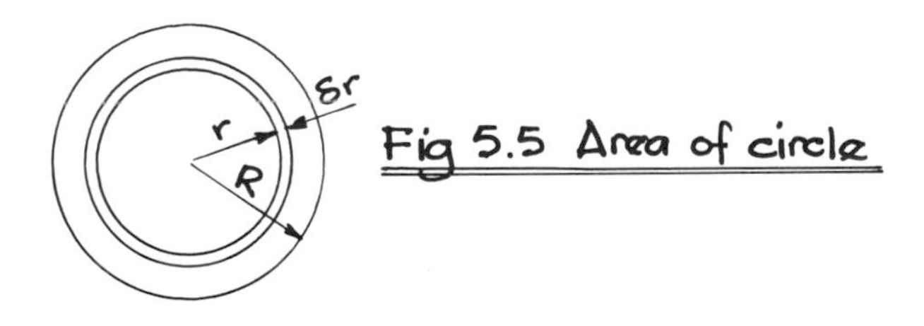

Fig 5.5 Area of circle

Knowing the circumference of the strip as $2\pi r$, its area is $2\pi r.\delta r$. If we now take the limits of r as 0 and R, then the area of the circle is $\int_0^R 2\pi r.dr$ which equals $\left[\frac{2\pi r^2}{2}\right]_0^R$. Therefore area = πR^2

CENTROIDS OF AREAS

We can also use integration to locate centroids of areas using the *first moment of area* technique as covered earlier in chapter 3. Figure 5.6 shows a section of graph (similar to that shown in fig. 5.3)

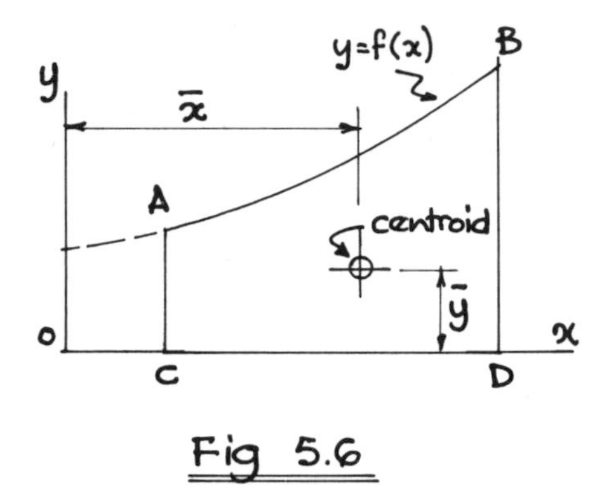

Fig 5.6

whose curve shows y as a function of x. It also indicates the centroid of its area in terms of $\bar{y}$ and $\bar{x}$. Figure 5.7 shows an elemental strip of width δx located x distance from AC and $y/2$ distance from CD. We can consider each strip to find its first moment of area and summate these to give a *total* first moment of area. If we then divide by the *total* moment, we can obtain the location of the centroid.

Stage 1: to find $\bar{y}$. Area of strip = $y.\delta x$. Area moment of strip = $y.\delta x \times y/2 = y^2 \delta x/2$. Total moment of area = $\int_A^B y^2 dx/2$. Total area (from previous example) = $\int_A^B y dx$.

Therefore $$\bar{y} = \frac{\tfrac{1}{2}\int_A^B y^2 dx}{\int_A^B y dx}$$

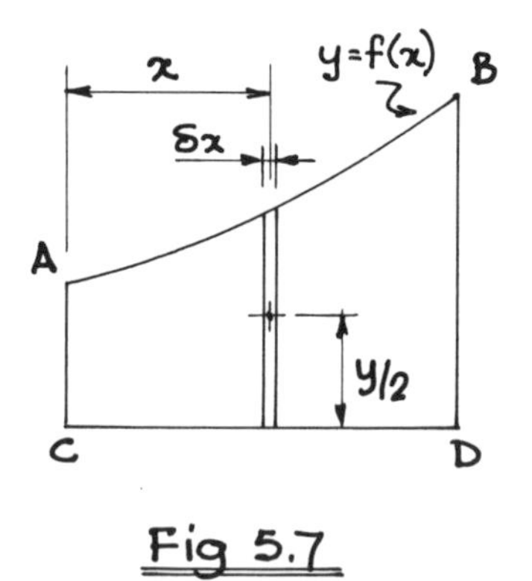

Fig 5.7

Stage 2: to find $\bar{x}$. Area of strip = $y.\delta x$. Area moment of strip = $x.y\delta x$

$$\text{Total moment of area} = \int_A^B x.ydx$$

$$\text{Total area} = \int_A^B y.dx$$

Therefore
$$\bar{x} = \frac{\int_A^B x.ydx}{\int_A^B y.dx}$$

Example 5.10
Figure 5.8 shows the parabolic curve $y = x^2 + 2x + 1$ between the limits of - 1 and 2 (as in example 5.8) but indicating the location of the centroid in terms of $\bar{x}$ and $\bar{y}$. Determine the location of its centroid and check the answer against the formula shown in Volume 1, chapter 8.

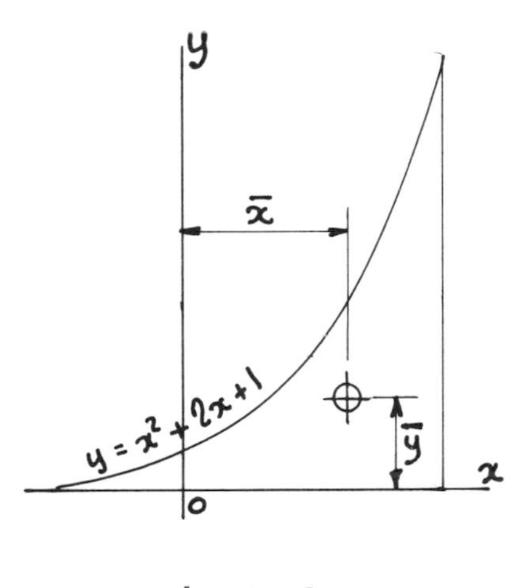

Fig 5.8

From the notes above,

$$\bar{y} = \frac{\tfrac{1}{2}\int_{-1}^{2}(x^2 + 2x + 1)^2\,dx}{\int_{-1}^{2}(x^2 + 2x + 1)dx}$$

In order to integrate the top of the fraction we must clear the brackets $(x^2 + 2x + 1)(x^2 + 2x + 1)$

$$= x^4 + 4x^3 + 6x^2 + 4x + 1$$

$$\tfrac{1}{2}\int_{-1}^{2} (x^4 + 4x^3 + 6x^2 + 4x + 1)dx$$

$$= \tfrac{1}{2}\left[\frac{x^5}{5} + x^4 + 2x^3 + 2x^2 + x\right]_{-1}^{2}$$

$$= \tfrac{1}{2}[(6.4 + 16 + 16 + 8 + 2) - (-0.2 + 1 - 2 + 2 - 1)]$$

$$= \tfrac{1}{2}[48.4 + 0.2] = 24.3$$

Area, from example 5.8, = 9

$$\bar{y} = 24.3 \div 9 = 2.7 \text{ up from X–X axis}$$

Also, from notes above,

$$\bar{x} = \frac{\int_{-1}^{2} x(x^2 + 2x + 1)dx}{\int_{-1}^{2} (x^2 + 2x + 1)dx}$$

In order to integrate the top line we clear the brackets:

$$\int_{-1}^{2} (x^3 + 2x^2 + x)dx$$

$$= \left[\frac{x^4}{4} + \frac{2x^3}{3} + \frac{x^2}{2}\right]_{-1}^{2}$$

$$= (4 + 5.33 + 2) - (0.25 - 0.67 + 0.5)$$

$$= 11.33 - 0.08 = 11.25$$

$$\bar{x} = 11.25 \div 9 = 1.25 \text{ to right of Y–Y axis}$$

(or 0.75 from right-hand side of graph).

From Volume 1, chapter 8,

$\bar{y} = 3/10 \times \text{height} = 0.3 \times 9 = 2.7$ up from base

$\bar{x} = 0.25$ from right-hand side $= 0.25 \times 3 = 0.75$

These confirm the figures.

SECOND MOMENTS OF AREA

We know, from chapter 3, that if we know where the neutral axis of a rectangle is, we can take *area moments squared* for each elemental strip to find the total moment of inertia (I), or second moment of area. A rectangular beam section of breadth b and depth d is shown in fig. 5.9. The neutral axis (X–X) is $d/2$ from

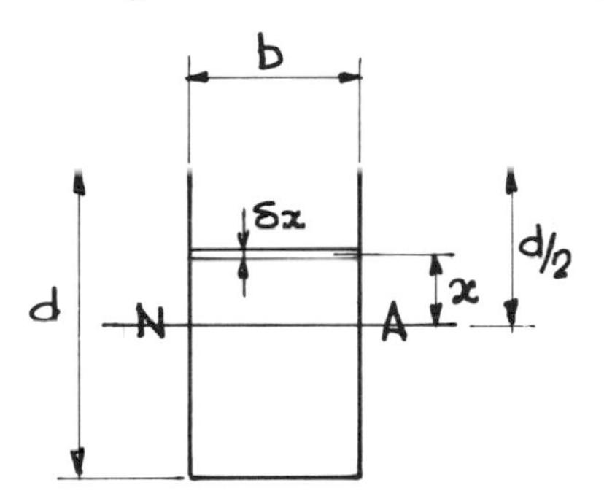

Fig 5.9

the top. The elemental strip $b.\delta x$ is x distance from the neutral axis so its second moment of area is $b.x^2\,\delta x$ so the summation about the neutral axis is $\int_0^{d/2} 2.bx^2\,dx$ (since there are as many strips *below* and *above* the axis).

$$I = 2b\left[\frac{x^3}{3}\right]_0^{d/2} = \frac{2bd^3}{3 \times 8} = \frac{bd^3}{12}$$

This proves the assumption made in chapter 3.

CASE STUDY 5.1 PROPPED CANTILEVER RETAINING WALL

Figure 5.10 shows the cross-section through a reinforced concrete basement wall. The ground-floor slab acts as a prop to support the top of the wall and resist its tendency to deflect under the force

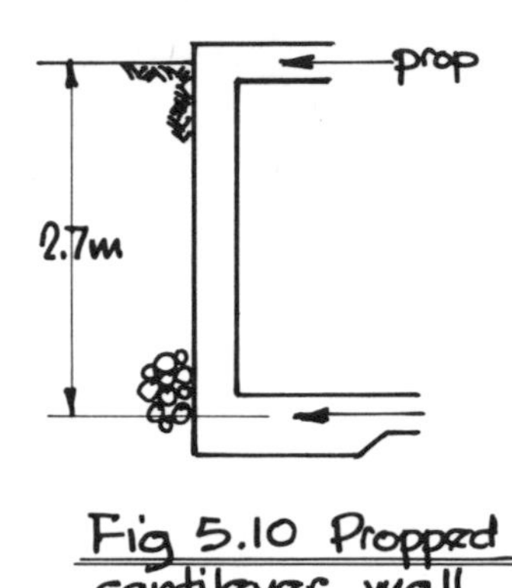

Fig 5.10 Propped cantilever wall

exerted by the earth pressure. The capacity of the ground-floor slab to perform this role is not questioned since it should provide adequate resistance to such movement *when it has gained sufficient strength*. Until that time, however, it will be necessary to resist the force by means of 'temporary works'.

In determining the bending effect on the propped wall, it is necessary to calculate the *required propping force* that will resist deflection. This is done by *superposition*. What this means in practice is that the wall is assumed initially to act as a cantilever, i.e. the propping force is removed. The triangular load produced by the earth pressure is then allowed to deflect the top of the wall to a measured movement (in terms of EI). Then a second condition is considered, where the earth pressure is removed, with the ground-floor slab acting as a point load at the top of the wall which cantilevers in the opposite direction and the point load is of the exact magnitude to deflect the wall precisely the same amount in the opposite direction (see fig. 5.11). Thus, when the conditions are superimposed, the wall does not move at its upper support. In order to avoid problems with variable ground water pressure, a system of land drainage is located near the base of the wall to carry the ground water around the basement. The density (ρ) of the retained earth = 1700 kg/m^3, and its angle of internal friction ϕ = 35°. In mathematical terms, in order to solve the equation we use trigonometry, moments of area, centroids of area and integral calculus (though we may use constants that we have already proved). The procedure is:

(a) determine the earth pressure at the base of the wall;

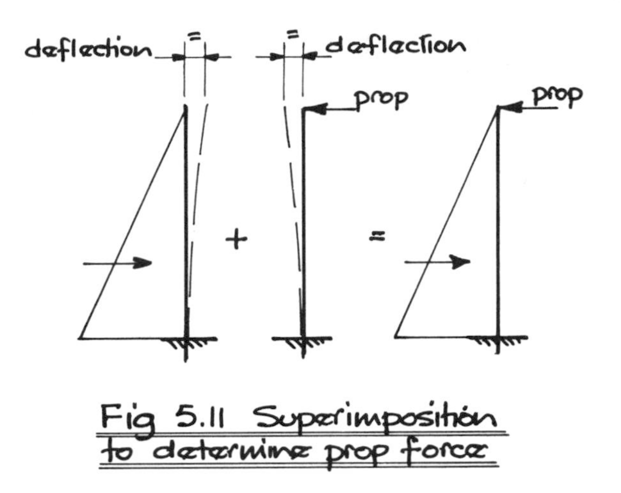

Fig 5.11 Superimposition to determine prop force

(b) determine the total earth pressure providing a force on the wall over a 1 m length of wall;
(c) determine the bending moment diagram for a cantilever condition (no prop);
(d) determine the resultant deflection at the top of the wall, (location shown in fig. 5.11) for the 'free' condition;
(e) determine the deflection caused by a point load P acting at the top of the wall in terms of P/EI;
(f) equate the deflections for both conditions to determine P
(g) determine reaction at base of wall.

Note: if we wished, we could continue from this stage to find the 'real' shearing force diagram, for the 'pinned' condition, and from this the bending moment maximum on the stem of the wall. However, this is outside the scope of National Certificate/Diploma studies and is best left to HNC.

Stage 1. Determine earth pressure at base of wall.

$$\text{Pressure at base} = \rho h \times \frac{1 - \sin\phi}{1 + \sin\phi}$$

$$= 17 \times 2.7 \times \frac{1 - \sin 35^\circ}{1 + \sin 35^\circ} \quad \text{(converting kg to kN)}$$

$$= 12.438 \text{ kN/m}$$

Stage 2. Determine total pressure and centre of action.

Force from ground pressure = area of triangle

$P_e = \frac{1}{2} \times 12.438 \times 2.7 = 16.791$ kN/m

Force P_e acts one-third of the height of retained earth = 0.9 m

Stage 3. Determine bending moment diagram. If we look at fig. 5.12, we can see that the bending moment at any point x can be determined from the expression $\frac{12.438x}{2.7} \times \frac{x}{2} \times \frac{x}{3} = 0.767\dot{7}x^3$

i.e. base x height/2 x lever arm (position of centroid). We can check this against the maximum bending moment of 16.791 x 0.9 = 15.11 kN m, or $0.767\dot{7} \times 2.7^3 = 15.11$ kN m.

Stage 4. Determine deflection for the top of the wall. To do this we need to find the area of the bending moment diagram and its centre of force. Using calculus, since we do not already know the area under a cubic graph or its centroid, we know the function of the bending moment curve is $y = 0.767\dot{7}x^3$

Therefore area under curve = $\int_0^{2.7} 0.767\dot{7}x^3 \, .dx$

$$= \left[\frac{0.767\dot{7}x^4}{4}\right]_0^{2.7} = \frac{0.767\dot{7} \times 2.7^4}{4}$$

$$= 10.2 \text{ kN m}^3$$

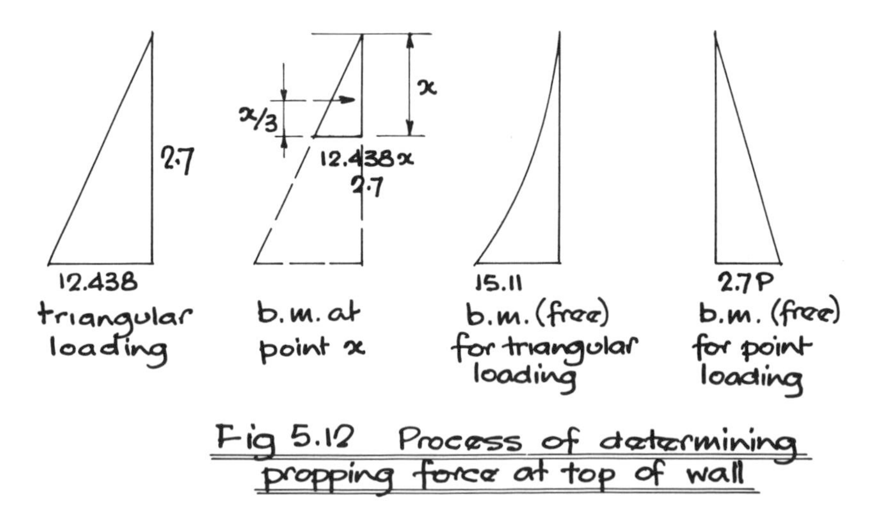

Fig 5.12 Process of determining propping force at top of wall

The centroid can be found using

$$\bar{x} = \frac{\int x.y\,dx}{\int y\,dx}$$

$$= \frac{\int_0^{2.7} 0.7677^{\cdot} x^4\,dx}{10.2}$$

$$= \left[\frac{0.7677^{\cdot} \times x^5}{10.2 \times 5}\right]_0^{2.7}$$

$$= 0.01505 \times 2.7^5$$

$$= 2.16 \text{ m from the top}$$

Therefore resultant deflection:

BM diagram x distance of centroid from maximum point of deflection

$$= \frac{10.2 \times 2.16}{EI} = \frac{22.03}{EI}$$

Note: this could also have been found by successive integration, which we will illustrate as a check on the result for stage 5.

Stage 5. Deflection from propping force P: BM = $P \times 2.7$ (force x distance for cantilever). BM diagram is triangular so area = $P \times 2.7^2/2 = 3.645P$. Centroid is two-thirds of distance from top of wall, so deflection = $(3.648P \times 1.8)/EI = 6.561P/EI$.

Check by successive integration: BM = Px. Slope = $\int$BM = $\int Px.\,dx = Px^2/2 + k$. Slope is zero at base of wall when $x = 2.7$, so $0 = (P \times 2.7^2)/2 + k - 7.29P/2 = k$. Slope = $P/2(x^2 - 7.29)$

$$\text{Deflection} = \int\text{Slope} = P/2\int(x^2 - 7.29)\,dx$$

$$= \frac{P}{2}\left[\frac{x^3}{3} - 7.29x + k\right]$$

Deflection is zero when $x = 2.7$, so $0 = \frac{2.7^3}{3} - 7.29 \times 2.7 + k$

$19.683 - 6.561 = k$

$k = 13.122$

$$\text{Deflection} = \frac{P}{2}\left[\frac{x^3}{3} - 7.29x + 13.122\right]$$

Deflection is maximum when $x = 0$ so maximum =

$$\frac{P}{2} \times \frac{13.122}{EI} = \frac{6.561P}{EI}$$

Stage 6. Equate deflections to find propping force:

$$\frac{6.561P}{EI} = \frac{22.03}{EI}$$

$P = 22.03 \div 6.561 = 3.36$ kN

Stage 7. Determine reaction at base: $\Sigma \rightarrow = \Sigma \leftarrow$ (horizontal forces equate to zero): 16.791 kN = 3.36 kN + reaction at base. Therefore reaction at base = 13.431 kN.

From this you can see that there are several ways to establish deflection when certain properties are known (e.g. we could also have used the coefficient for deflection for the point load of $PL^3/3EI$) but it is always sensible to use the most simple method. In this example, area moment is the most straightforward, and it is worth remembering that the area of the bending-moment diagram and its centroid will be constant for a triangular load of the configuration shown. Area of BM = ¼ base x height. Centroid from top = ⅕ x height. It is also worth noting the effect of the prop on the bending moment at the base of the wall, i.e. free BM for cantilever *minus* bending moment of prop, which equals 16.791 x 0.9 - 3.36 x 2.7 = 15.112 - 9.072 = 6.04 kNm. This represents a reduction in bending of about 60%, which illustrates the economic benefit to be gained. The effect is shown in fig. 5.13.

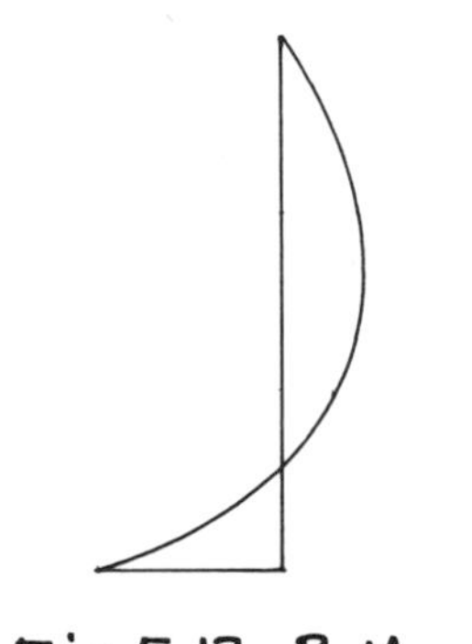

Fig 5.13 B.M. diagram for propped cantilever

Exercises

5.1 Integrate with respect to x: (i) x^3; (ii) $\sqrt{x}$; (iii) $1/x^2$.

5.2 Integrate with respect to a: (i) $3a-2$; (ii) $(a-4)(a-1)$; (iii) $(a+2)^2$.

5.3 Integrate with respect to b: (i) $6b^4 + 3b^2 - 2b + 1$; (ii) $b^2 + 1/b^2 - 3b/2$; (iii) $3(2b+1)$.

5.4 Evaluate the following definite integrals: (i) $\int_2^3 x.dx$; (ii) $\int_1^4 (2x-1)dx$; (iii) $\int_1^3 \sqrt{x}.dx$.

5.5 Evaluate the following definite integrals:
(i) $\int_1^2 (2x^2 - 4x + 2)\,dx$; (ii) $\int_2^4 \left(\frac{3}{x^2} + \frac{2}{x} + 1\right) dx$.

5.6 Calculate the area under the curve between the limits $x = 1$ and $x = 5$ for the equation $y = x^2 - 2x + 1$ for the graph shown in fig. 5.14.

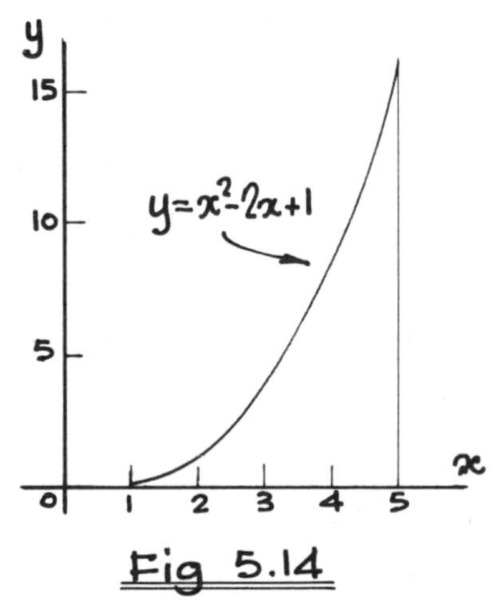

Fig 5.14

5.7 Determine the area under the curve $y = x^2 - 4x + 3$ between the limits $x = 1$ and $x = 3$ and its centroid of area about the x and y axes of the graph as shown in fig. 5.15.

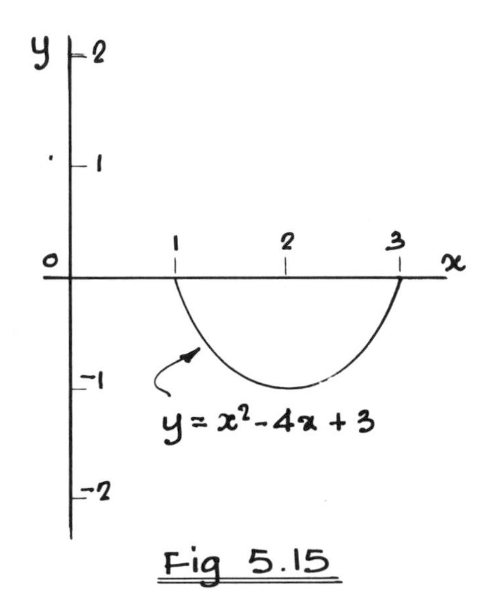

Fig 5.15

5.8 Determine, by successive integration, the deflection coefficient for a cantilever beam with uniform loading as shown in fig. 5.16.

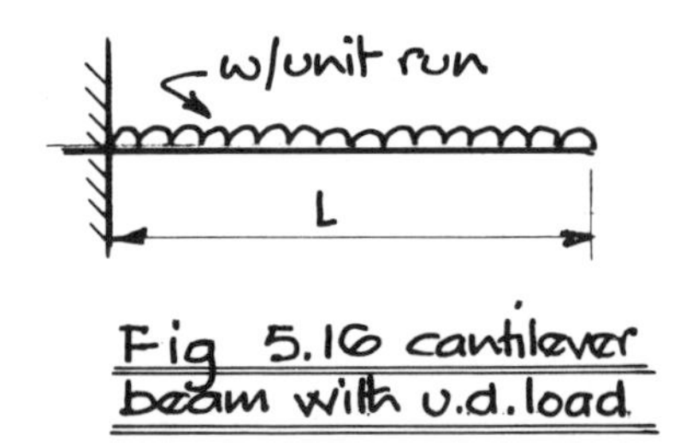

Fig 5.16 cantilever beam with u.d.load

5.9 Using the coefficient $PL^3/3EI$, as proved in the case study, determine the reactions for a propped cantilever beam subjected to uniform loading.

5.10 Using the answers obtained from questions 5.8 and 5.9, determine the location and magnitude of maximum span and support bending moments for a beam of propped cantilever form subjected to uniform loading of rate w per metre run.

6 Statistics: Data Handling

Statistics is one of the most commonly employed branches of mathematics in general everyday use. In part this is because it is an easy way of presenting information, either numerically or graphically. The way in which information (data) is collected, assembled and presented is known as 'data handling' or 'data management'. One way in which data is handled is by 'sampling' and translating the results of the sample into statistical information which can be used to promote an idea. Alternatively, the information may be used to predict a pattern of events.

Statistics is also popular because it does not require the precision of other areas of mathematics (such as Trigonometry), but it must be realised that this lack of precision can produce quite odd results if it is not based on truly representative sampling within reasonable limits of acceptability. Unfortunately, statistics can often be abused and distorted to give the impression of something other than what is really happening, and this is particularly true in the case of the graphical presentation of data. Several graphical techniques were illustrated in Volume 1, and it is easy to see that if scales are distorted, the resultant diagram can give a misleading visual impression of the information shown.

STATISTICAL TERMS

Statistics has only been developed seriously since the 1950s, and as with any 'new' discipline has produced its own language. The following is a summary of some of the more common terms which need to be understood in order to be able to use statistical data:

(1) *Population:* this is an entire set of results, e.g. if we carry out concrete cube tests at intervals in a contract, the population is the total rather than a selection.

(2) *Sample:* this is a small random section of the population and is intended to be representative of the population as a whole.
(3) *Continuous data:* this describes results that can be of any value, e.g. the results of concrete cube tests can be of any value.
(4) *Discrete data:* this describes information which has a finite value, e.g. the number of windows in a building.
(5) *Frequency polygon:* a curve formed by joining the mid-points of each 'class' in a histogram (see chapter 4, Volume 1).
(6) *Cumulative frequency:* the addition of all the frequencies to show the variable as a running total.
(7) *Relative frequency:* the 'class' frequency divided by the total frequency. Therefore the sum of all the relative frequencies is equal to unity.
(8) *Relative percentage frequency:* relative frequency expressed as a percentage rather than a fraction.
(9) *The mean:* the sum of all the parts in a 'set' divided by the number of parts in the 'set'.
(10) *The median:* the middle of a 'set'.
(11) *The mode:* the number in the 'set' which occurs most frequently.
(12) *The ogive:* the shape of the curve achieved by plotting a cumulative frequency.
(13) *Quartiles:* the division of cumulative frequencies into four equal parts.
(14) *Percentiles:* the division of cumulative frequencies into a hundred equal parts.

PRESENTATION OF DATA

In Volume 1 we considered alternative methods of displaying data and it is worth reviewing some of these here. Suppose a local authority employs the following number of craftsmen in its 'direct works' department:

Type of employee	Number employed	Percentage of total
Bricklayers	96	40
Carpenters	36	15
Electricians	36	15
Gas fitters	24	10
Plumbers	48	20
Total	240	100

This information can be represented pictorially in the ways shown in fig. 6.1.

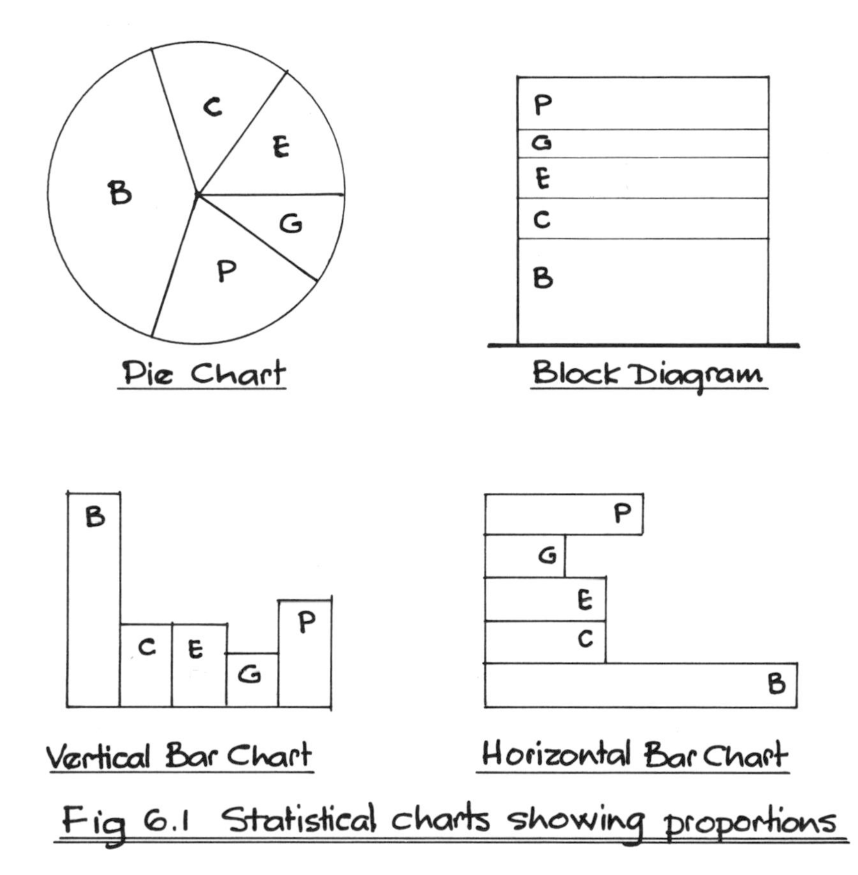

Fig 6.1 Statistical charts showing proportions

The pie chart: indicates the proportion as a 'share of the pie', which can be understood very easily.

The block diagram: relies on heights, or areas, to convey the proportions, the overall height being 100%.

The bar chart: which can show, either vertically or horizontally, a comparison of the various components by the length of the bars. Although providing a good comparison of components, it does not provide easy access to the total number of employees.

FREQUENCY DISTRIBUTIONS

In quality control we may wish to know the frequency with which a component matches an exact specification and the frequency with which, or number of times, it varies by a measured amount of similar values. In simple terms we use two processes to produce a graphical representation: (1) a system of tally marks to count the frequency, (2) a histogram to convert these data to a more graphical format.

Example 6.1
Fifty reconstructed stone blocks of nominal length 220 mm are measured and the results tabulated as below:

221	219	220	223	221	217	218	220
217	220	222	221	220	220	221	
220	220	221	220	220	221	220	
222	220	220	221	220	223	220	
217	218	220	219	220	223	222	
222	222	219	218	219	220	223	
222	220	217	222	221	219	221	

Stage 1. Construct a tally of the various different sizes from least length (217) to greatest length (223) starting at the top left-hand corner and reached left to right:

217	||||	4
218	|||	3
219	~~||||~~	5
220*	~~||||~~ ~~||||~~ ~~||||~~ |||	18
221	~~||||~~ ||||	9
222	~~||||~~ ||	7
223	||||	4

*Target length.

The frequency distribution can now be presented as a histogram to give a clearer visual representation, as shown in fig. 6.2.

Continuous data
When dealing with continuous data it is often useful to group the information into classes or categories. If we take a typical example of concrete cube test results, we can see how such data can be grouped.

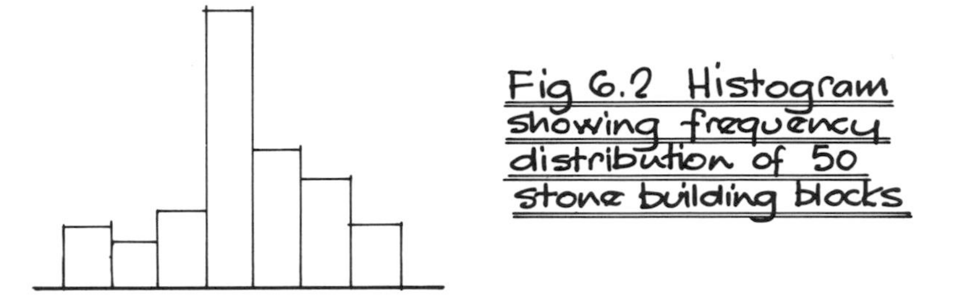

Fig 6.2 Histogram showing frequency distribution of 50 stone building blocks

Example 6.2

The results of compressive strength tests on 30 concrete test cubes are listed as follows (in N/mm^2):

26.0	24.0	21.6	23.2	29.4
24.6	26.7	26.7	25.5	25.8
26.8	27.1	25.7	28.7	27.2
25.6	26.2	25.6	26.0	28.3
24.2	22.0	26.4	27.2	27.4
26.3	27.7	28.8	30.0	28.5

In order to group the data, we take the upper and lower values to give us the range, i.e. 30−21.6 = 8.4. This is known as the *class range*, with 30 and 21.6 being the *class limits.* The class size is determined by taking the class range and dividing it into equal parts. *Note:* the number of parts should divide easily into the total number of cubes *and* the range. Both 30 and 8.4 are divisible by 6, so there are six classes each with a range of 1.4. As before we use a tally but in this case it is using *grouped data.* To avoid an overlap of classes, we start below the lowest value and finish above the highest, thus:

Compressive stress	No. of cubes	Frequency
21.4–22.8	||	2
22.9–24.3	|||	3
24.4–25.8	卌 |	6
25.9–27.3	卌 卌 |	11
27.4–28.8	卌 |	6
28.9–30.3	||	2

As in the previous example, we can represent this information on a histogram indicating the middle points of each block (see fig. 6.3).

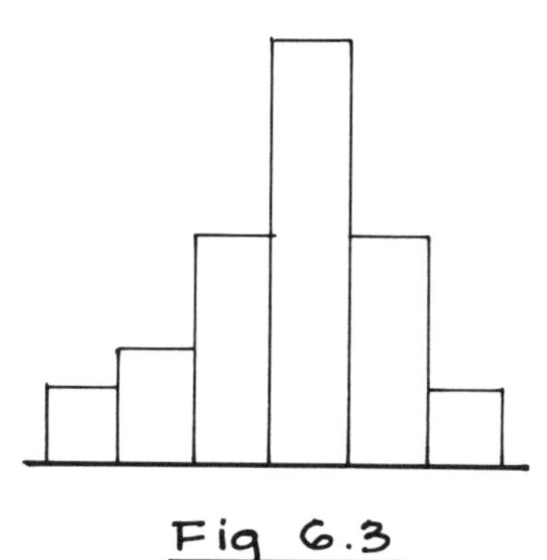

Fig 6.3

Sometimes we may wish to look at part of the distribution in greater detail, notably at that with the greatest frequency, and in such cases we can narrow the classes towards the middle of the range and extend them at the limits. For example, if we consider 100 cube tests rather than 30, we might extend the first and last group to a range of 4.0 N/mm^2, the second and last but one to a range of 2.0 N/mm^2, and the remainder to a range of 1.0 N/mm^2, giving the following frequencies:

Compressive stress	Frequency
18.5–22.5	4
22.6–24.6	8
24.7–25.7	18
25.8–26.8	36
26.9–27.9	20
28.0–30.0	10
30.1–34.1	4

When drawing the histogram, however, it must be remembered that it is the *area* of the rectangle that represents the frequency, and not the *height*. This can be seen on the histogram in fig. 6.4. (Since the classes vary in range, the scale has been omitted from the vertical axis.)

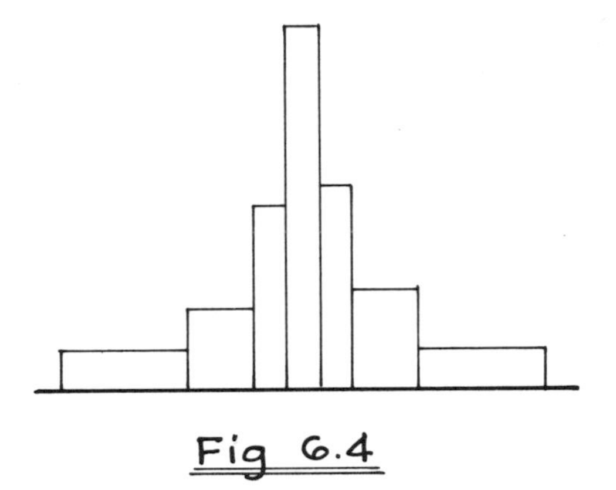

Fig 6.4

The frequency polygon

As explained at the start of the chapter, and shown in Volume 1, the frequency polygon is obtained by joining the mid-points of the tops of the grouped data frequencies in the histogram. This is shown in fig. 6.5, which converts the histograms of figs 6.2 to 6.4 into frequency polygons. As can be seen, these are all line graphs (i.e. a series of straight lines joining the mid-points). However, as the

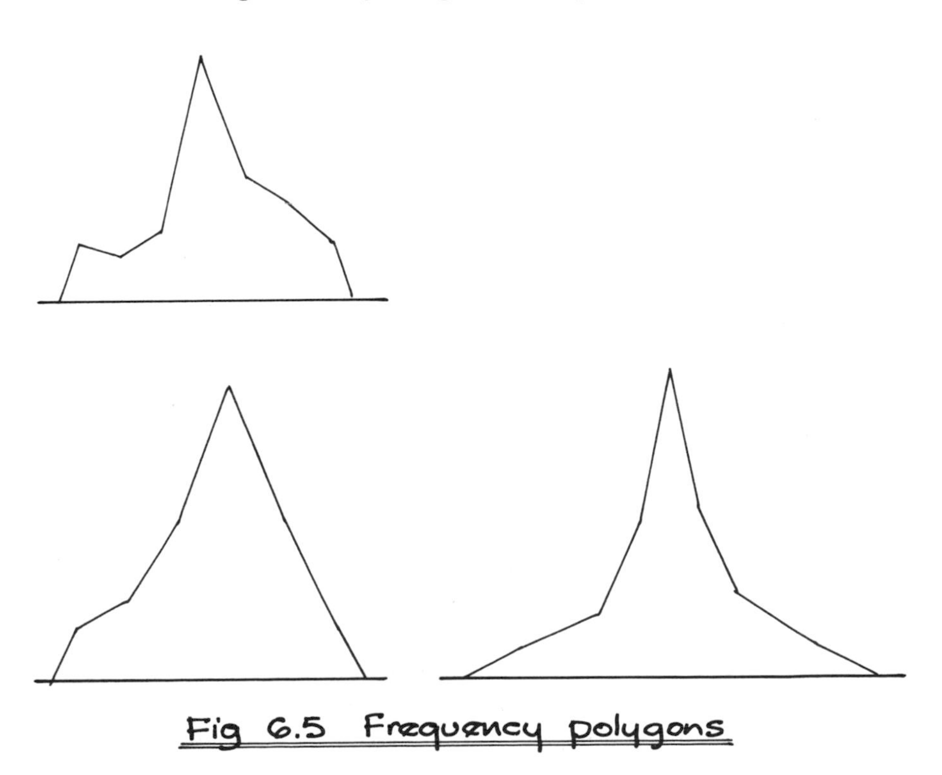

Fig 6.5 Frequency polygons

number of results increases and the class sizes reduce, the line graphs become much closer to a smooth curve. This smoothed curve tends to be a truer representation of the total population, and, since some curves can be represented by equations, certain characteristics can be evolved without the need to refer to a set of data. One example of the curve described above is the 'normal distribution curve', which we will consider in chapter 7.

CUMULATIVE FREQUENCY

This is an alternative method of representing frequency distributions and is obtained by taking the lower class boundaries and adding their frequencies as a running total. Let us reconsider the reconstructed stone blocks examined in example 6.1. If we take the length values as being the middle of each rectangle on the histogram, each being 1 mm width, we can construct a second table thus:

Length of block as measured	Cumulative frequency
less than 216.5	0
less than 217.5	0 + 4 = 4
less than 218.5	4 + 3 = 7
less than 219.5	7 + 5 = 12
less than 220.5	12 + 18 = 30
less than 221.5	30 + 9 = 39
less than 222.5	39 + 7 = 46
less than 223.5	46 + 4 = 50

When plotted as a cumulative frequency graph the curve follows the form of an 'ogive', which is an architectural term for a profile curve. This is seen in fig. 6.6.

RELATIVE FREQUENCY

We may encounter many situations where the actual frequency is less important than the relative frequency. The relative frequency of a class is determined from the expression 'relative frequency = class frequency/total frequency'. If we again consider the reconstructed stone blocks from example 6.1, the measurements, their frequency and their relative frequency can be tabulated as follows:

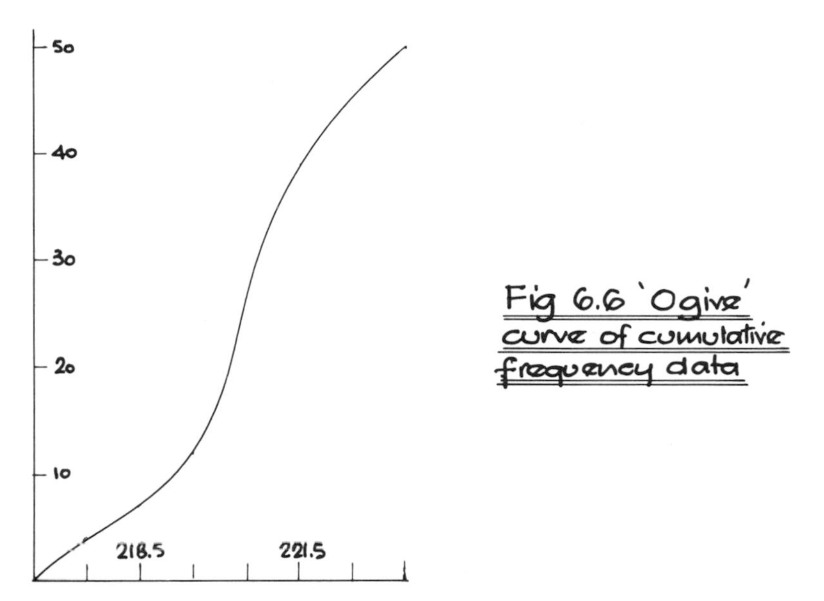

Fig 6.6 'Ogive' curve of cumulative frequency data

Width of measured sample	Frequency	Relative frequency
217	4	4/50 = 0.08
218	3	3/50 = 0.06
219	5	5/50 = 0.10
220	18	18/50 = 0.36
221	9	9/50 = 0.18
222	7	7/50 = 0.14
223	4	4/50 = 0.08
Total	50	1.00

Alternatively it may be more informative to convert the fractions to percentages rather than decimals, since people are not as likely to look at the frequency (50) when discussing the result, and are more likely to look at percentages instead. Rather than produce another variation of the results table shown above, the following examples illustrate how the relative frequencies may be converted quite rapidly, e.g. 5/50 = 10%, and 18/50 = 36%. The polygons for both the relative frequency and the relative percentage frequency have the same shape, the only variation being the scale of the vertical axis (see fig. 6.7).

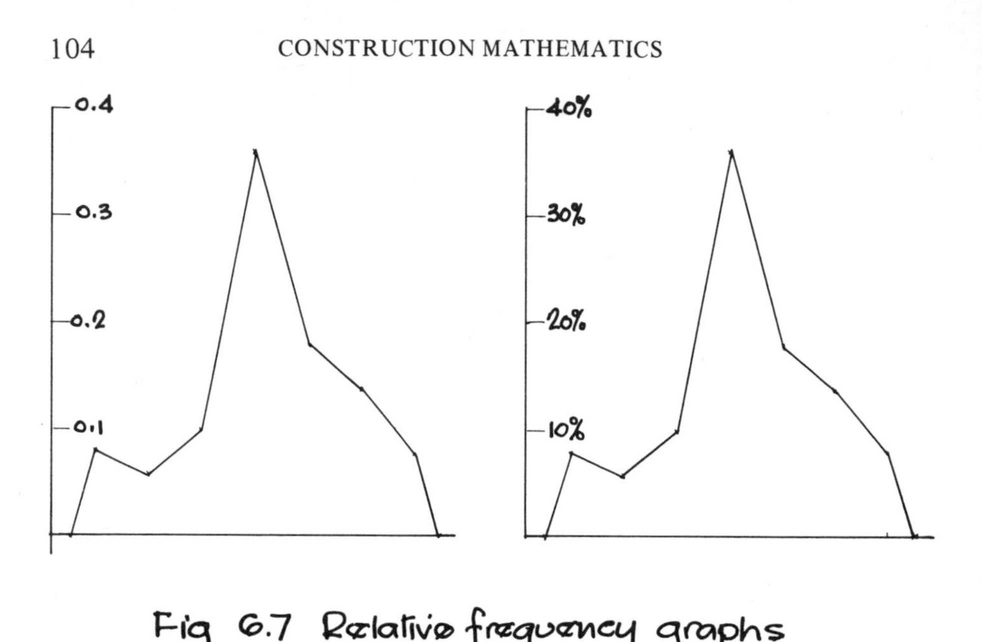

Fig 6.7 Relative frequency graphs

MEAN, MEDIAN AND MODE

These terms tend to be thought of collectively and are probably the best known of all statistical terms.

The mean

The arithmetic mean of a set of values is determined by adding all the values together and then dividing by the number of values in the set, i.e. arithmetic mean = summation of values/number of values.

Example 6.3

Ten concrete test cubes are tested to destruction at 7 days after casting. Their compressive stresses are found to be: 23.6, 28.2, 24.3, 27.2, 27.1, 26.4, 28.9, 31.0, 24.4, 30.6 (all values are in N/mm^2). Calculate the mean compressive stress of the set.

Mean =

$$\frac{23.6 + 28.2 + 24.3 + 27.2 + 27.1 + 26.4 + 28.9 + 31.0 + 24.4 + 30.6}{10}$$

$$= 27.17\ N/mm^2$$

Alternatively, since we can see by inspection that the result should be around 27, we can base our summation on the variations from that, i.e.

$$\frac{-3.4 + 1.2 - 2.7 + 0.2 + 0.1 - 0.6 + 1.9 + 4.0 - 2.6 + 3.6}{10}$$

$$= 0.17 + 27 = 27.17 \text{ N/mm}^2$$

Note: if you are not this confident, you can, nevertheless, reduce the calculation by varying from 20, e.g. 3.6 + 8.2 + 4.3, etc., which will produce 71.7 ÷ 10 = 7.17 + 20 = 27.17 N/mm².

If we have already started a statistical analysis of a set of values and have produced a frequency distribution, it is not necessary to go back to the beginning and summate all the values independently. Knowing the frequency with which some of the values occur, we simply summate the products of value x frequency and then divide by the total. For instance, in example 6.1 we considered 50 reconstructed stone blocks of varying lengths. Having found that there were just seven variations, it is obviously easier to use this information. To do this we call the lengths x_1, x_2, x_3, x_4, x_5, x_6, x_7, and the frequencies with which they occur f_1, f_2, f_3, f_4, f_5, f_6, f_7. We then summate $x_1 f_1$, $x_2 f_2$, etc., which can be represented by the expression:

$$\text{mean } \bar{x} = \frac{\Sigma xf}{\Sigma f}$$

which, for this example, equals

$$\frac{(217 \times 4)+(218 \times 3)+(219 \times 5)+(220 \times 18)+(221 \times 9)+(222 \times 7)+(223 \times 4)}{4 + 3 + 5 + 18 + 9 + 7 + 4}$$

$$= \frac{868 + 654 + 1095 + 3960 + 1989 + 1554 + 892}{50}$$

$$= 11012 \div 50 = 220.24 \text{ mm}$$

As explained in the previous example, since we already know the 'target length' is 220 mm, we could have simplified the calculation to:

$$\frac{(-3 \times 4)+(-2 \times 3)+(-1 \times 5)+(0 \times 18)+(1 \times 9)+(2 \times 7)+(3 \times 4)}{50}$$

$$= \frac{-12-6-5+0+9+14+12}{50} = \frac{12}{50} = 0.24$$

$$\text{mean} = 220 + 0.24 = 220.24$$

Note: this probably illustrates this technique better than it did in example 6.3 since the calculation can be done *easily* without a calculator.

If, however, we have 'grouped' the data in classes, we must use the value of the *mid-point* of each class to represent x.

Example 6.4

Using the values obtained for cube test results in example 6.2, determine the mean of the grouped distribution. Since we know that most cubes fail around 26.6 N/mm², we can use this as a base for the stresses (above or below) to simplify the calculation. This is best tabulated, thus:

Compressive stress	Mid-point	Coded value	Frequency	Coded result
21.4 → 22.8	22.1	− 4.5	2	− 9
22.9 → 24.3	23.6	− 3.0	3	− 9
24.4 → 25.8	25.1	− 1.5	6	− 9
25.9 → 27.3	26.6	0	11	0
27.4 → 28.8	28.1	+ 1.5	6	+ 9
28.9 → 30.3	29.6	+ 3.0	2	+ 6

$$x_c = \frac{\Sigma x_c f}{\Sigma f} = \frac{-12}{30} = -0.4$$

Therefore: $\text{mean} = 26.6 - 0.4 = 26.2\ \text{N/mm}^2$

The median

The median is, quite simply, the middle value of a set of numbers or values (or a distribution) where the data is in 'ranked order'. This means that the values must be in either ascending or descending order. If the set of data has an odd number of values, then the median is the middle value. If, however, the set has an even number of values, then the median is the mean of the two middle values, e.g. the median for 1, 2, 2, 3, 3, 4, **5**, 5, 6, 6, 7, 8, 8 is 5, whereas for 1, 2, 2, 3, **4**, **4**, 5, 5, 6, 7 the median is (4 + 4)/2, which equals 4. When we consider a distribution for which we have produced a frequency polygon or histogram, then the median is the middle point of the diagram since this 'peak' is where most values occur and is in ranked order. Thus the median for the reconstructed stone blocks considered in example 6.1 is 220 mm, and for the 100 concrete cube test results analysed in the early part of the chapter it is 26.3 N/mm^2.

The mode

The mode of a set of numbers or values is that which occurs most frequently. Thus the mode of 2, 3, 3, 4, 4, 4, 5, 6 is 4, since 4 occurs most frequently. This type of sequence is known as *unimodal*, since it has only *one* mode. If we look at another sequence, 8, 8, 7, 7, 7, 6, 6, 5, 5, 4, 4, 4, 3, 3, 2, 2, there are *two* modes (7 and 4), since both numbers occur with the same frequency. This type of sequence is called *bimodal*. In many instances, however, there may be *no* mode, e.g. 1, 2, 3, 4, 5, 6, 7, 8, 9 has no value occurring more frequently than any of the others. In order to find the mode of a frequency distribution, we look at the histogram, which indicates the mode immediately as its highest point.

QUARTILES AND PERCENTILES

We saw earlier that the median divides a set of values into two equal parts. The *quartile* divides the set into *four* equal parts and these are usually denoted by Q_1, Q_2, Q_3 where Q_2 is also the median.

Example 6.5

Find the first, second and third quartiles for the following set of numbers: 1, 3, 6, 8, 5, 9, 4, 10, 5, 3, 7, 2.

We must first 'rank' the set of numbers, which, for convention's sake, we will do in ascending order, thus: 1, 2, 3, 3, 4, 5, 5, 6, 7, 8, 9, 10. Since there are twelve numbers in the set, the first quartile (Q_1) falls between the third and fourth numbers, i.e. $Q_1 = (3 + 3)/2 = 3$, the second quartile (Q_2) falls between the sixth and seventh numbers, i.e. $Q_2 = (5 + 5)/2 = 5$, and the third quartile falls between the ninth and tenth numbers, i.e. $Q_3 = (7 + 8)/2 = 7.5$. Thus $Q_1 = 3$, $Q_2 = 5$, $Q_3 = 7.5$. (*Note:* for a frequency distribution the quartiles correspond to ¼, ½ and ¾ of the total frequency. The quartiles can, therefore, be represented on an ogive.)

Having seen that the median divides a set of values into *two* equal parts and a quartile divides the set into *four* equal parts, we should also note that the *percentiles* divide the set of values into a *hundred* equal parts. We only need to consider percentiles, therefore, when dealing with very large sets of values.

Exercises

6.1 A manufacturer of sealed-unit double-glazed windows carries out a laboratory test on a sample of 40 windows to check their U values. The results obtained are as follows:

2.80	2.90	2.85	2.95	2.80	2.90	3.05	2.90
2.85	2.95	2.90	3.00	2.75	2.95	3.10	2.95
2.80	2.85	2.85	3.10	2.70	3.05	3.00	2.85
2.90	2.75	3.00	2.95	2.80	2.95	2.95	2.90
2.95	2.85	3.05	2.90	2.85	3.10	3.00	2.80

Determine the frequency distribution for the sample and produce a histogram to illustrate the results.

6.2 Prior to delivery, a quota of timber floor joists is checked at the timber yard for Young's modulus properties by passing samples through a deflection meter. The computed sample results are, in N/mm^2 :

10 300	10 500	11 200	10 700	12 100	11 800	11 500
11 100	10 100	10 200	8 800	9 100	10 800	10 000
10 700	9 500	10 000	8 700	10 200	10 300	9 600
9 900	9 300	9 500	9 200	9 700	9 900	9 800
10 000	8 800	9 000	8 700	8 500	9 000	9 400
9 600	9 400	9 200	8 800	9 000	9 600	9 800

Using grouped data, produce a histogram to indicate the frequency distribution.

6.3 Using the data produced from exercise 6.1, develop a cumulative frequency table and graph for the U values.

6.4 Using the data produced from exercise 6.2, determine the relative frequency of E values and produce a relative percentage frequency graph.

6.5 A mechanical excavator operator, working under fairly consistent conditions, carries out a series of trench excavations over a period of 15 working days. The amount of excavation measured each day is, in metres cubed:

36.6	42.0	38.4	37.5	40.5
39.2	38.0	38.6	40.0	40.2
37.5	37.0	38.8	41.5	39.6

Establish the mean excavation rate and the median rate so that an estimate of acceptable daily rate can be made for future contracts.

7 Statistics: Distributions and Probabilities

We saw in chapter 6 that if we take a large enough sample we obtain a frequency distribution curve which is balanced about its centre line. This is known as a *normal distribution* curve, as illustrated in Volume 1. Five other types of distribution curve are shown in fig. 7.1. These are:

(a) a positive skewed curve;
(b) a negative skewed curve;
(c) a U-shaped curve, which normally occurs when a minimum value is being sought;
(d) and (e) defective data curves, which normally occur when frequency distributions of a number of defective items are plotted. For example, if 100 clocks were to be checked against the BBC 'pips', we would expect some to be slow, some to be fast and some to be correct. If, however, 30 of them were governed by a master control, a false distribution would be shown.

The tendency of a distribution is given by the mean, median and mode, but two different distributions may have the same mean with a very different spread of dispersion (see fig. 7.2). One way to assess the likely spread of dispersion is from examining the *range of distribution*, where range = largest observation - smallest observation.

Example 7.1
Ten steel cables of similar cross-section sustain maximum loads (in kilonewtons) as listed:

110.4	89.3	88.4	109.6	87.2
113.4	100.2	101.4	101.3	102.7

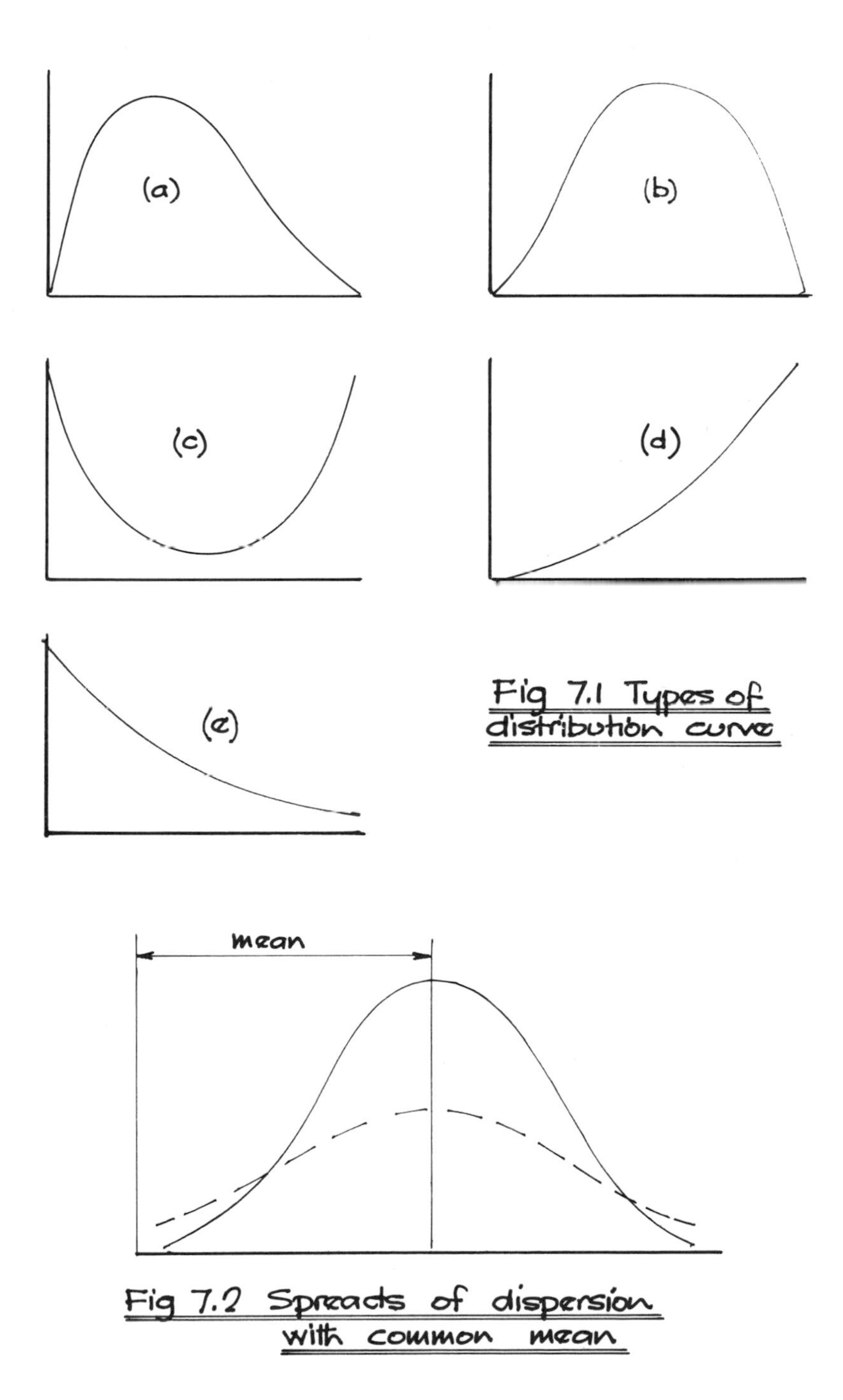

Fig 7.1 Types of distribution curve

Fig 7.2 Spreads of dispersion with common mean

Determine the range and the mean.

The range = 113.4 - 87.2 = 26.2 kN

The mean, using a base of 100, equals

$$\frac{10.4 - 10.7 - 11.6 + 9.6 - 12.8 + 13.4 + 0.2 + 1.4 + 1.3 + 2.7}{10}$$

$$= +3.9 \div 10 = +0.39 \qquad \text{mean} = 100.39 \text{ kN}$$

The range, together with the mean, gives us a fair idea of the distribution, but it does depend on the maximum and minimum values. Also, in some cases it is difficult to calculate the range. We therefore need something to allow us to compare the spread of two different curves. The measure we use is called the *standard deviation.*

STANDARD DEVIATION

The standard deviation (σ) is a deviation from the arithmetic mean expressed as a *root mean square*, thus:

$$\sigma = \sqrt{\frac{\Sigma x^2 f}{\Sigma f} - \bar{x}^2} \quad \text{or} \quad \sqrt{\frac{\Sigma f(x - \bar{x})^2}{\Sigma f}}$$

where $\bar{x}$ = the arithmetic mean, x = the variable, and f = the frequency.

Example 7.2

A sample of 100 engineering bricks is tested to determine the compressive stress. A frequency table of the test results is given below:

Stress (N/mm^2)	24.7	25.0	25.3	25.6	25.9
Frequency	5	18	50	20	7

Calculate the mean and the standard deviation of the frequency distribution.

Since we know that *mean* $\bar{x} = \dfrac{\Sigma xf}{\Sigma f}$

and *standard deviation* $\sigma = \sqrt{\frac{\Sigma x^2 f}{\Sigma f} - \bar{x}^2}$

we can tabulate the information as follows:

x	x^2	f	xf	$x^2 f$
24.7	610.09	5	123.5	3 050.45
25.0	625.00	18	450.0	11 250.00
25.3	640.09	50	1 265.0	32 004.50
25.6	655.36	20	512.0	13 107.20
25.9	670.81	7	181.3	4 695.67
Total		100	2 531.8	64 107.82

$$\bar{x} = \frac{\Sigma xf}{\Sigma f} = \frac{2531.8}{100} = 25.318 \text{ N/mm}^2$$

$$\sigma = \sqrt{\frac{\Sigma x^2 f}{\Sigma f} - \bar{x}^2} = \sqrt{\frac{64107.82}{100} - 641.001}$$

$$= \sqrt{641.0782 - 641.001} = \sqrt{0.0771}$$

$$\sigma = 0.278 \text{ N/mm}^2$$

When considering a grouped distribution, the standard deviation is calculated by taking x as the mid-points of the class widths.

Example 7.3
The number of bricks laid per bricklayer per day for a sample of 75 bricklayers is as follows:

No. of bricks	600–640	641–681	682–722	723–763	764–804	805–845
Frequency	3	6	19	27	18	2

We can tabulate in a similar way to example 7.2, remembering that x is the *mid-point* of each class width.

No. of bricks	x	f	x^2	xf	x^2f
600–640	620	3	384 400	1 860	1 153 200
641–681	661	6	436 921	3 966	2 621 526
682–722	702	19	492 804	13 338	9 363 276
723–763	743	27	552 049	20 061	14 905 323
764–804	784	18	614 656	14 112	11 063 808
805–845	825	2	680 625	1 650	1 361 250
Total		75		54 987	40 468 383

$$\bar{x} = \frac{\Sigma xf}{\Sigma f} = \frac{54\,987}{75} = 733.16 \text{ bricks}$$

$$\bar{x}^2 = 537\,523.59$$

$$\sigma = \sqrt{\frac{\Sigma x^2 f}{\Sigma f} - \bar{x}^2} = \sqrt{\frac{40\,468\,383}{75} - 537\,523.59}$$

$$= \sqrt{539\,578.44 - 537\,523.59} = \sqrt{2054.85}$$

$$= 45.33 \text{ bricks}$$

Answer

Mean (per bricklayer) = 733 bricks
Standard deviation = 45 bricks

THE NORMAL CURVE

We have discussed normal distribution and the normal curve. Being able to determine standard deviation and the mean allows us to relate all normal curves. The standard deviation is a measure of the spread of the curve about its mean. For practical purposes the normal curve is assumed to terminate at three standard deviations on either side of the mean (theoretically the normal curve will extend to infinity).

It can be calculated that the area under a normal curve as a percentage of the total area is the same for all normal curves at one, two and three standard deviations. The area between one standard deviation either side of the mean ($\bar{x} \pm 1\sigma$) is approximately 68%. The area between two standard deviations on either side of the mean ($\bar{x} \pm 2\sigma$) is 95%, and the area between three standard deviations either side of the mean ($\bar{x} \pm 3\sigma$) is virtually 100% of the total area under the curve (see fig. 7.3).

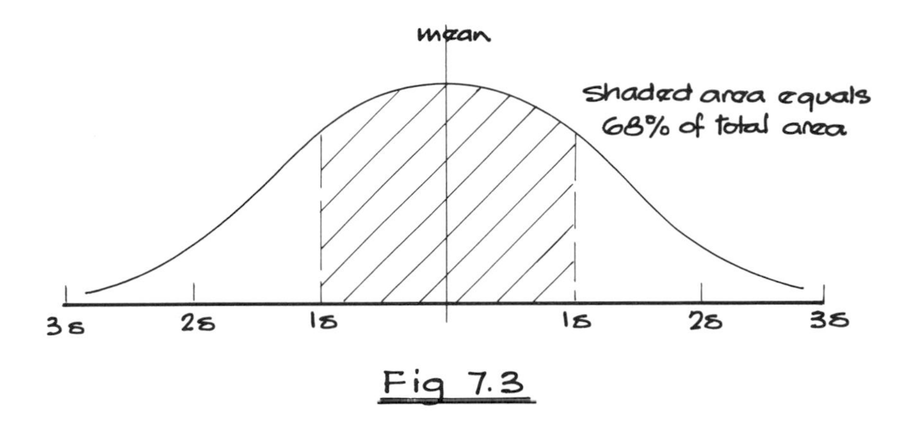

Fig 7.3

Example 7.4
One hundred concrete cubes are tested after 7 days. The results are tabulated below. How many of the cubes developed a compressive stress in excess of 26.42 N/mm^2?

Compressive stress (N/mm^2)	22.0	23.5	25.0	26.5	28.0
Frequency	5	25	40	25	5

In order to answer the question, we need to establish the mean and the standard deviation for the distribution.

x	f	x^2	fx	fx^2
22.0	5	484	110	2 420
23.5	25	552.25	587.5	13 806.25
25.0	40	625	1 000	25 000
26.5	25	702.25	662.5	17 556.25
28.0	5	784	140	3 920
Total	100		2 500	62 702.5

$$\bar{x} = \frac{\Sigma xf}{\Sigma f} = \frac{2500}{100} = 25 \text{ N/mm}^2$$

$$\bar{x}^2 = 625$$

$$\sigma = \sqrt{\frac{\Sigma x^2 f}{\Sigma f} - \bar{x}^2} = \sqrt{\frac{62\ 702.5}{100} - 625}$$

$$\sigma = \sqrt{627.025 - 625} = \sqrt{2.025}$$

$$\sigma = 1.42 \text{ N/mm}^2$$

From this we can see that $\bar{x} + 1\sigma = 26.42$ N/mm² and, since we know that $\bar{x} \pm 1\sigma = 68\%$ of total, this means that $32\% \div 2 = 16\%$ to either side of normal distribution. Therefore 16% of the cubes exceeded 26.42 N/mm² under test.

In practice, of course, we are concerned with how many cubes fall *within* a prescribed range rather than outside it. In order to reduce the risk of weak concrete occurring in a structure, we select 90% of the sample as being the range. The standard deviation for this is $\pm\ 1.64\sigma$ from the mean. So, for the example given above, the 'characteristic' strength of the concrete is *mean* $-$ *1.64*σ which equals $25 - 1.64 \times 1.42 = 25 - 2.33 = 22.67$ N/mm².

Logically we should also assess loadings statistically but there is often insufficient data available to produce sufficiently accurate results and we use empirical loadings for different situations. In the case of wind loads, however, a statistical approach is adopted to take account of the designed 'life' of the structure, e.g. the likelihood of wind speeds reaching 150 km/h are greater in a 100-year span than in a 20-year span, and in America many road bridges are designed for a 20-year life because the traffic flow and capacity will demand that they be replaced by that time. This leads us into the consideration of probabilities.

PROBABILITY

Every time we are asked to make a choice, such as calling 'heads or tails' when a coin is spun, or sampling from a batch of components because we cannot test each one to destruction, we are involved with probability. With the coin the probability is that it will come down the way we call, and with the component the probability is that it will be representative of the sample and not lie outside the range of acceptability.

Probability is a measure of what, given every possibility, the chance is of something happening or not happening. The sum of the 'happenings' and 'not happenings' is always one.

Simple probability

If we take the example of tossing the coin, there are only two possibilities, either heads or tails, since:

$$\text{probability } (p) = \frac{\text{number of ways in which an event can happen}}{\text{total number of possibilities}}$$

thus $p_{\text{heads}} = 1/2$ and $p_{\text{tails}} = 1/2$.

A slightly more complex example is the rolling of a die. Since there are six sides, the probability of rolling a six (or any other specified number) is 1/6.

Total probability

Having explained that the sum of the 'happenings' and 'not happenings' is one, we can see that whereas the probability of rolling a die to score a 6 is 1/6, the probability of instead scoring a 1, 2, 3, 4 or 5 (not a 6) is 5/6 so that the total probability = 1/6 + 5/6 = 1.

Mutually exclusive events

If we draw a card from a pack of playing cards, it will be either 'red' or 'black' but not both. These types of event are called mutually exclusive, since not more than one of them can occur. If, however, we choose between an ace and a black card, we might draw a black ace, so these events are *not* mutually exclusive.

Addition of probabilities

If two or more events are mutually exclusive, the total probability is the sum of the parts.

Example 7.5

What is the chance of cutting an ace, a king, a queen or a jack in a single cut of a pack of cards?

$p_{ace} = 4/52 = 1/13$ $\quad$ $p_{king} = 4/52 = 1/13$

$p_{queen} = 4/52 = 1/13$ $\quad$ $p_{jack} = 4/52 = 1/13$

$$p_{ace + king + queen + jack} = 1/13 + 1/13 + 1/13 + 1/13 = 4/13 = 0.308$$

We can see, therefore, that, where events are mutually exclusive, $p_{total} = p_1 + p_2 + p_3 + \ldots p_n$

DEPENDENT AND INDEPENDENT EVENTS

The spinning of a coin is an example of an independent event. Events are *independent* when the outcome of one event has no effect on the probability of the outcome of the next event, i.e. if a coin lands heads on one spin, it has no effect on the probability of it landing either heads or tails on its next spin. The probability of drawing an ace from a pack of cards is completely independent of the probability of drawing a king, but if we do not replace the ace, then we have changed the probability of drawing a king or another ace. That is, $p_{ace} = 4/52 = 1/13$; $p_{king} = 4/52 = 1/13$, but p_{king} (after ace has been drawn) = 4/51, and p_{ace} (after ace has been drawn) = 3/51 = 1/17. These are called *dependent* events, and it can be seen that whereas the probability of a king has increased marginally, the possibility of another ace has reduced considerably.

Multiplication of probabilities

When two or more events are independent, the probability of their both happening is determined by multiplying the probability of each together, i.e. $p = p_1 \times p_2 \times p_3$, etc. For example, if we know that the probability of drawing either an ace or a king from a full pack of cards is 1/13, then the probability of drawing an ace, replacing it, and then drawing a king is 1/13 x 1/13 = 1/169.

Combination of addition and multiplication of probabilities

It is possible to have a mixture of independent and dependent events. If we know that 90% of concrete cubes fall within an acceptable range of results, this implies that 5% are above and 5% below this range. If, therefore, we take three results at random, then the probability of all three being below the acceptable range is:

$$p_{all\ three} = p_1 \times p_2 \times p_3 = 0.05 \times 0.05 \times 0.05 = 0.000125$$

because all three events are independent. To determine the possibility of only one defective cube in the three tested, we have three combinations:

1st	*2nd*	*3rd*
good	good	fail
good	fail	good
fail	good	good

Any one of these three gives only one defective part as they are mutually exclusive events. We know that 95% of the cubes are not defective (including those that fall outside the upper limit of the range), so the probability will be:

$$(0.95 \times 0.95 \times 0.05) + (0.95 \times 0.05 \times 0.95) + (0.05 \times 0.95 \times 0.95)$$

$$= 0.045 + 0.045 + 0.045 = 0.135$$

We have concerned ourselves primarily with the *normal* distribution so far, but two other distributions used commonly in statistics are the *binomial* and *Poisson* distributions. In order to understand the binomial distribution, we must be familiar with binomial expansions, which requires us to return briefly to algebra.

BINOMIAL EXPANSION

A binomial is the sum or difference of two terms such as: $(a + b)$, $(2 - x)$, $(3b - 2c)$, etc. It is often necessary to expand (multiply out) a binomial which is raised to a power, e.g. $(2 + x)^2 = (2 + x)(2 + x) = 4 + 4x + x^2$. This is quite straightforward for small powers, and was covered in Volume 1, but for larger powers, such as $(a + b)^6$, the calculation could be quite tedious. A useful device for helping us to ascertain the numerical components of the expansion is Pascal's triangle.

Pascal's triangle

Let us consider the expression $(1 + x)$

$$(1 + x)^0 = 1$$
$$(1 + x)^1 = 1 + x$$
$$(1 + x)^2 = 1 + 2x + x^2$$

$$(1+x)^3 = 1 + 3x + 3x^2 + x^3$$
$$(1+x)^4 = 1 + 4x + 6x^2 + 4x^3 + x^4$$
$$(1+x)^5 = 1 + 5x + 10x^2 + 10x^3 + 5x^4 + x^5$$
$$(1+x)^6 = 1 + 6x + 15x^2 + 20x^3 + 15x^4 + 6x^5 + x^6$$

Now let us take the numerical component of each term as listed above, e.g. for $20x^3$ take 20 and arrange the numbers in the shape of an isosceles triangle:

$$\begin{array}{ll}
(1+x)^0 & 1 \\
(1+x)^1 & 1 \quad 1 \\
(1+x)^2 & 1 \quad 2 \quad 1 \\
(1+x)^3 & 1 \quad 3 \quad 3 \quad 1 \\
(1+x)^4 & 1 \quad 4 \quad 6 \quad 4 \quad 1 \\
(1+x)^5 & 1 \quad 5 \quad 10 \quad 10 \quad 5 \quad 1 \\
(1+x)^6 & 1 \quad 6 \quad 15 \quad 20 \quad 15 \quad 6 \quad 1
\end{array}$$

When the numbers are arranged in this way, it is known as *Pascal's triangle*. This has certain features:

(1) The numerical components for the first and last terms are always 1, i.e. the outside of the triangle is always 1.
(2) Each component in the triangle is obtained by adding the two components above it, e.g. $3 + 3 = 6$ and $1 + 5 = 6$.
(3) The number of terms in the expansion is always one more than the index power, e.g. $(1 + x)^3$ has four terms.

Although Pascal's triangle gives the coefficients, it is not very practical to use it for very large indices and a theorem is used as follows:

$$(a+b)^n = a^n + na^{n-1}b + \frac{n(n-1)a^{n-2}b^2}{2!} + \frac{n(n-1)(n-2)a^{n-3}b^3}{3!}, \text{ etc.}$$

The expression 3! means 3 factorial or $3 \times 2 \times 1$. Similarly, 7! means $7 \times 6 \times 5 \times 4 \times 3 \times 2 \times 1$, and so on.

This still requires a fair amount of effort, but if the second of the binomials is very small, you will see that its additional contribution to the expansion is very small. For example, when, in the expression $(1 + x)^4$, x equals 0.01, then $(1 + x)^4$ would expand to:

$$1^4 + 4 \times 1^3 \times 0.01 + \frac{4(4-1)1^2 \times 0.01^2}{2 \times 1} + \frac{4(4-1)(4-2)1^1 \times 0.01^3}{3 \times 2 \times 1} +$$

$$\frac{4(4-1)(4-2)(4-3)1^0 \times 0.01^4}{4 \times 3 \times 2 \times 1}$$

which simplifies to: 1 + 0.04 + 0.0006 + 0.000004 + 0.00000001 = 1.04060401. We can see that from the third term onwards the values become insignificant. From this observation we can say that when x is very small, then $(1 + x)^n \simeq 1 + nx$ or $(1 - x)^n \simeq 1 - nx$. This is known as the *binomial approximation* and can appear in a number of forms, e.g.

(1) $\sqrt{1+x} = (1+x)^{1/2} \simeq 1 + x/2$

(2) $\sqrt{a+x} = a^{1/2}(1 + x/a)^{1/2} \simeq a^{1/2}(1 + x/2a)$

(3) $\dfrac{1}{1+x} = (1+x)^{-1} \simeq 1 - x$

(4) $\dfrac{1}{a-x} = (a-x)^{-1} = a^{-1}(1 - x/a)^{-1} \simeq a^{-1}(1 + x/a)$

$$= \frac{1 + x/a}{a} = \frac{a+x}{a^2} = \frac{1}{a} + \frac{x}{a^2}$$

As an example of its application, consider the value 1/4.95. Obviously this can be solved using a calculator, i.e. 0.20202, but if we make $a = 5$ and $x = 0.05$, the value 1/4.95 can be written as

$$\frac{1}{a-x} = \frac{1}{a} + \frac{x}{a^2} = \frac{1}{5} + \frac{0.05}{25}$$

$$= 0.2 + 0.002 = 0.202$$

which has the same accuracy as the calculated result to four significant figures! Remember, though, this approximation is valid only for small values of x.

BINOMIAL DISTRIBUTION

Returning to statistics, earlier in the chapter we discussed the addition of probabilities such as $(p + q)$, which is a binomial expression. The total probability of $(p + q)$ can be expanded as $(p + q)^n$ to give a binomial distribution from which a theoretical frequency table can be drawn. A histogram can then be drawn to show the distribution.

Example 7.6

Calculate the probability, and express the result as a histogram, that in a sample of five concrete test cubes either 0, 1, 2, 3, 4 or 5 will exceed the minimum if only one of five is allowed to fail.

Probability of failure $q = 0.2$
Probability of success $p = 1 - 0.2 = 0.8$

$(p + q)^n$ expresses *all* the probabilities where n is the sample size which equals 5. From our knowledge of the binomial expansion $(p + q)^5$ equals:

$$p^5 + 5p^4q + 10p^3q^2 + 10p^2q^3 + 5pq^4 + q^5$$

which, when tabulated, produces:

No. of defectives	Term in expansion	Probability	
0	p^5	$(0.8)^5$	= 0.32768
1	$5p^4q$	$5(0.8)^4(0.2)$	= 0.40960
2	$10p^3q^2$	$10(0.8)^3(0.2)^2$	= 0.20480
3	$10p^2q^3$	$10(0.8)^2(0.2)^3$	= 0.05120
4	$5pq^4$	$5(0.8)(0.2)^4$	= 0.00640
5	q^5	$(0.2)^5$	= 0.00032
Total probability			1.00000

The results are shown on a histogram in fig. 7.4.

MEAN AND STANDARD DEVIATION OF THE BINOMIAL DISTRIBUTION

In chapter 6 we saw that the mean:

$$\bar{x} = \frac{\Sigma xf}{\Sigma f}$$

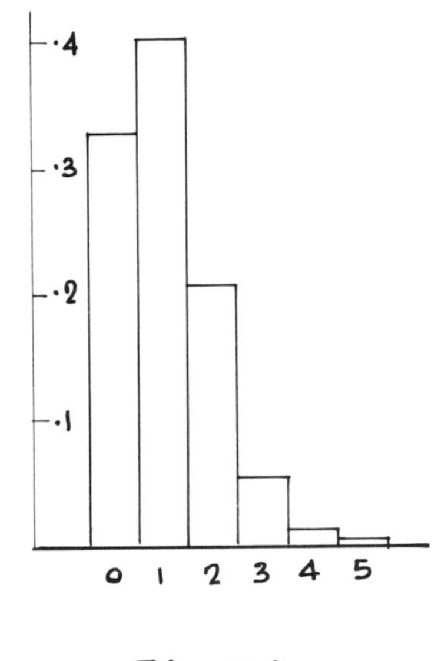

Fig 7.4

and the standard deviation:

$$\sigma = \sqrt{\frac{\Sigma f(x - \bar{x})^2}{\Sigma f}}$$

We saw also that the *mean* and *standard deviation* of a frequency distribution could be found by using the appropriate formula. The mean and standard deviation of a binomial distribution can be determined similarly using the formulae:

Mean number of successes $= np$

With a standard deviation $= \sqrt{npq}$

these being derived from the basic formulae.

Example 7.7

In a sample of five concrete test cubes taken from a batch which has a 1 in 20 failure rate, what would be the expected number of good cubes, and what is the standard deviation?

Total of all cubes $= p + q$

$q = 1/20 = 0.05$

$p = 1 - 0.05 = 0.95$

Mean $= np = 5 \times 0.95 = 4.75$

Standard deviation $= \sqrt{5 \times 0.95 \times 0.05}$

$= 0.487$

POISSON DISTRIBUTION

The Poisson distribution is a special form of the binomial distribution. The binomial distribution may be used in all cases to calculate the probabilities, but for large samples (n) the mathematics involved becomes very long. In the construction industry, as explained in *Construction Science* Volume 1, a high degree of precision is not a realistic requirement in calculating probabilities, and it is possible to use an approximation of the binomial distribution. The Poisson distribution is one example of this and can be used when:

(a) the sample size is greater than 10;
(b) the probability of the events happening $p \leqslant 0.10$;
(c) the mean number (np) is less than 5.

If all the above conditions are satisfied, the Poisson distribution can be represented by:

$$\text{probability } [x] = \frac{e^{-c} c^x}{x!}$$

where $c = np$ (the expectation of it happening)
$x =$ number of happenings
$e =$ the base of natural logarithms $= 2.7183$

Values of e^{-c} are given in most tables and can also be obtained from a scientific calculator.

Example 7.8

Of all the garages constructed by a builder, 10% need some form of repair within 1 year of completion. Of 20 garages built, what is the probability of 0, 1, 2, 3, 4 or 5 garages (in a smaller sample) needing repair?

$c = np$ where $n = 20, p = 0.1$

$c = 20 \times 0.1 = 20$

$$p[0] = \frac{e^{-2} \times 2^0}{0!} = 0.1353$$

(*Note:* this is an expression for a probability *not* an equation so 0! is acceptable.)

$$p[1] = \frac{e^{-2} \times 2^1}{1!} = 0.2706$$

$$p[2] = \frac{e^{-2} \times 2^2}{2!} = 0.2706$$

$$p[3] = \frac{e^{-2} \times 2^3}{3!} = 0.1804$$

$$p[4] = \frac{e^{-2} \times 2^4}{4!} = 0.0902$$

$$p[5] = \frac{e^{-2} \times 2^5}{5!} = 0.0361$$

The histogram for this example can be seen in fig. 7.5.

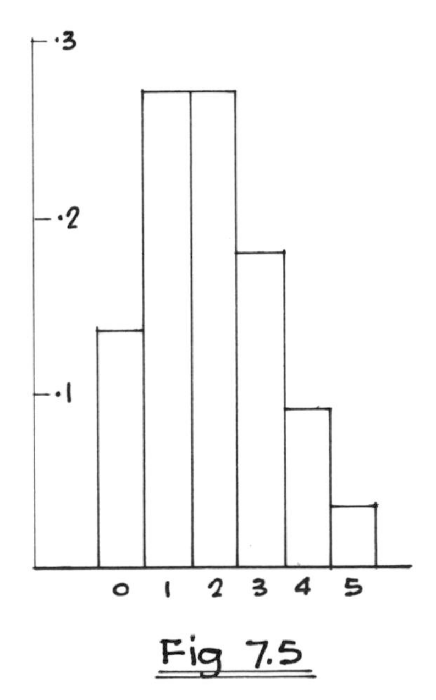

Fig 7.5

Exercises

7.1 Determine the standard deviation for the U value of the sample of windows considered in chapter 6 (exercise 6.1) and the 'characteristic' value for heat-loss calculation purposes if $U_c = \bar{x} - 1.5\sigma$.

7.2 Calculate the characteristic strength for the concrete tested in chapter 6 (example 6.2) if the acceptable range of the sample is 90%.

7.3 The timber joists considered in chapter 6 (exercise 6.2) fall within an acceptable range of 95%. If we take four joists at random, what are the probabilities of (a) all four joists, (b) one of the four joists, falling below the 'characteristic' modulus of elasticity? Check how many, if any, actually fall below this figure.

7.4 Find the first three terms of $(x - 6y)^{10}$ and determine the approximate value of the full expansion if $x = 8$ and $y = 0.01$.

7.5 Plot the binomial distribution of the expression $(a + b)^6$, given $a = 0.9$ and $b = 0.1$, and determine its standard deviation.

8 Computer Applications

In the last thirty years we have witnessed a quite remarkable development in the area of computers. This to a large extent has been harnessed to the development in microelectronics, which in turn was a spin-off from the space programme where every gram of excess load was critical. In the mid 1950s, computers were physically very large and, because of their mass and volume, were located at the lowest level of a building, e.g. the ground floor or basement. They were also very sensitive and required a dust-free atmosphere, which meant providing air locks to fire-escape exits. An example of how the equipment has developed is the modern desktop micro, which is both more efficient than the 1950s monster described above and less sensitive in that it does not require a special environment.

For all its bulk, the early computer was really quite limited in its applications, being confined mainly to arithmetical processes and data storage. It is in these fields of application that it is best known even today, but these are now by no means the limits of its commercial applications. Some typical examples of its relevance to the construction industry are:

Architecture

two-dimensional and three-dimensional graphics including perspectives (CAD);
storage of performance data for materials, components and equipment;
scheduling of windows, doors, etc.;
specification writing and updating (word processing).

Building production

work planning by arrow/precedence networks and critical-path analysis;
progress logging through bar charts;

inventories of materials, machinery, etc.;
financing, including forecasts of cash flow, turnover and profitability.

Civil engineering
draughting/detailing and scheduling allied to structural work (computer-aided draughting);
analysis and design in a variety of areas;
surveying through electronic measurement, plotting and calculation (total survey station).

Environmental engineering
calculations related to thermal, acoustic and illumination problems;
fluid mechanics including sizing for pipes, ducts and open channels.

Quantity surveying
estimating, costing, measurement, data banks;
bills of quantities, specifications.

Valuation of property
data storage and retrieval;
financial forecasting;
scheduling of dilapidations;
report writing.

TERMS IN COMMON USE

Computers fall into four main categories, *mainframe, supermini, mini,* and *micro,* and most firms now have at least one of these. Computing has unfortunately developed a jargon of its own, even in relation to the prime components of computers, and it is necessary to know some of the basic terms in order to understand how they function.

The *hardware* is what makes the computer work, and comprises a *central processor*, which is fed by *input* devices. The most common input is the *keyboard*, which is an extension of the typewriter layout. A typical keyboard is shown on the front cover of this book. A second input device, popular on some machines, is a *card reader*, and *rigid tapes* are used for hand-held computers in association with a simplified keyboard. Graphics systems may use *joysticks* or a *stylus* and *pixel* board. So that we can use the computer, we must be able to read an *output*, which can be either temporary, for intermediate stages of work, or permanent, so that it can be filed for future use. The most usual temporary output is the *visual display unit* (VDU), which is effectively a small television monitor screen in either monochrome or colour; the latter is particularly useful for graphics applications. Permanent output, or *hard copy*, can be produced by a *dot matrix* printer, *daisy wheel* printer or *laser* printer for lettering, or by a *plotter* for draughting work.

Software is the information that feeds into the central processor, and in order for it to be effective it must be presented in a logical way. The process of arranging this information is called a *program*, which may be written by the operator or as a commercial package for general use. A benefit of the latter is that it has already been tried and tested before its sale in order to iron out any snags (*debugging*), whereas in writing a program from scratch there is a fair amount of trial and error involved.

All computers have a built-in *memory* (random access memory, RAM) which can be used for storing data. Unfortunately this only works while the machine is switched on. Once it has been switched off, this memory is lost unless it is stored on a secondary storage system. Mainframe computers use either magnetic *reel-to-reel tapes* or *rigid discs*, and microcomputers use *cassettes* or *floppy discs*. The tape or cassette stores information in sequence and has to be run through to find the specific data required, but a disc system 'dumps' information and can access it more quickly. Most personal business microcomputers therefore have *disc drive*, but a large proportion of home computers use cassette storage. A simple diagrammatic layout of a microcomputer is shown in fig. 8.1.

Obviously it is not possible to cover all the applications that have been described in just one chapter, and we shall confine ourselves to arithmetic applications appropriate to the mathematics covered in Volume 1 and 2.

PROGRAMMABLE CALCULATORS

In Volume 1 we introduced the scientific calculator and, since many of these have a programmable facility, it might be useful to illustrate a very simple program using a calculator.

Example 8.1
Program for determining annual mortgage repayments based on:

$$\text{pmt} = \frac{P \times i}{1 - (1 + i)^{-n}}$$ (see Volume 1, p. 89)

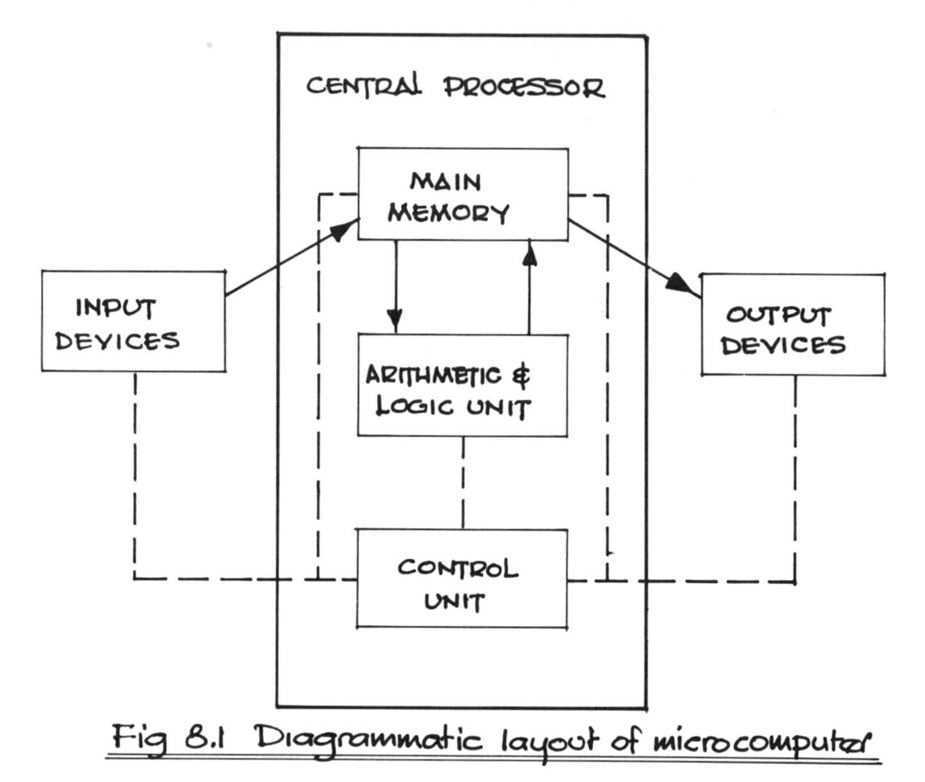

Fig 8.1 Diagrammatic layout of microcomputer

This allows us to examine alternative interest rates and years of mortgage, and also capital (though most repayment rates are quoted per £1000 by banks, building societies and insurance companies).

In order to write the program, we need to establish the right sequence of events, since the memory capacity of calculators is usually limited to 24 steps. The sequence is:

(1) deal with the bracket $(1 + i)$

(2) raise to power of n $(1 + i)^n$

(3) produce reciprocal $(1 + i)^{-n}$

(4) change to minus $-(1 + i)^{-n}$

(5) add the unit value $1 - (1 + i)^{-n}$

(6) produce reciprocal $\dfrac{1}{1 - (1 + i)^{-n}}$

(7) multiply by principal $\dfrac{P}{1 - (1 + i)^{-n}}$

(8) multiply by interest $\dfrac{P \times i}{1 - (1 + i)^{-n}}$

Each step can now be put, in sequence, into the learning program, but *remember* we must allow for the stages at which we produce a calculation by using 'equals'. Also we must prefix the program with the command 'learn' (LRN) and conclude it similarly. Each value (i, n, P, pmt) must be allowed for by using the 'stop' function in the learning program, thus:

	LRN		
00	1		
01	+	11	+
02	STOP(i)	12	1
03	=	13	=
04	log	14	$1/x$
05	x	15	x
06	STOP(n)	16	STOP(P)
07	=	17	x
08	10^x	18	STOP(i)
09	$1/x$	19	=
10	+/-	20	STOP(pmt)
			LRN

Remember this program is only stored while the calculator remains *on*. To use the program we can now run the sequence inserting the required values of i, n, P and i again to establish pmt. This can be illustrated in full for a £30 000 mortgage borrowed over 30 years at an interest rate per annum of 10% where $i = 0.10$ (% to decimal) $n = 25$ and $P = 30\,000$. To start the program we must go to the first step (00) so we press GOTO 00 then

00	RUN		
01	↓	06	Enter 25
02	Enter 0.10	07	RUN
03	RUN	08	↓
04	↓	09	↓
05	↓	10	↓

11	16	Enter 30 000
12	17	RUN
13	18	Enter 0.10
14	19	RUN
15 ↓	20	Result = 3305.04

Now, by clearing the entry and returning to the start through GOTO 00, we can obtain alternative payments thus:

$i = 0.12, n = 25, P = 30\,000$ result = 3825.00
$i = 0.13, n = 25, P = 30\,000$ result = 4092.78
$i = 0.13, n = 20, P = 30\,000$ result = 4270.61
$i = 0.10, n = 30, P = 30\,000$ result = 3182.38

These show that the time span does not have a great effect on repayments. For example, extending the time to 30 years for the 10% interest rate only reduces repayments by £122.66 p.a. (or £10.22 per month), and for the 13% interest rate a reduction of 5 years (from 25 to 20) only increases the repayments by £177.83 p.a. (or £14.82 per month). However, an increase in interest rate from 10% to 12% adds £519.96 p.a. (£43.33 per month) and from 10% to 13% adds £787.74 p.a. (£65.65 per month).

Returning to the computer, the sequences described above are developed into a *flow chart*. As with the example using the programmable calculator, each step, however obvious, has to be built in. There are two reasons for this: (1) the computer cannot think but can only follow a logical progression of events, (2) the program may be run by a computer operator who has only keyboard skills and is unfamiliar with the problem being solved.

FLOW CHARTS

Among the symbols commonly used in flowcharting are the following:

This shape of box normally contains the instruction to START or STOP the program.

This is the instruction box containing a description of the operation to be carried out, e.g. A = B + C or sum $x = 0$.

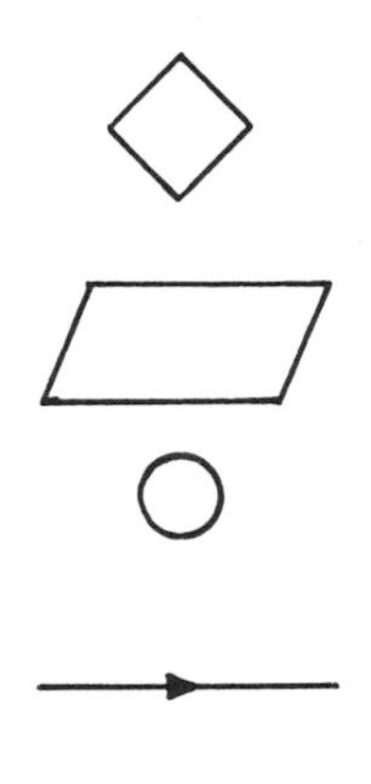

This is the question box, which usually requires a YES or NO answer allowing decision points and a choice of path.

This is the READ or PRINT box, which allows for information to be absorbed or for OUTPUT.

In order to allow the flow chart to be contained on one piece of paper this gives a junction which is coded – usually by a number.

The boxes must be connected by straight lines showing the direction of flow.

This process is illustrated in fig. 8.2, which shows a simple flow chart for calculating the average of five numbers.

Example 8.2
From chapter 7 we know that the standard deviation of compressive stress for a sample of 100 concrete test cubes can be found by using

$$\sigma = \sqrt{\frac{\Sigma x^2 f}{\Sigma f} - \bar{x}^2}$$

Alternatively this may be written as:

$$\sigma = \sqrt{\left(\frac{x^2 + x_1^2 + x_2^2 + x_3^2 + \ldots x_n^2}{n}\right) - \left(\frac{x + x_1 + x_2 + x_3 + \ldots x_n}{n}\right)^2}$$

if we consider each sample separately, rather than in sets of frequency, with n being the number of cubes.

We can now construct a flow chart to solve the values for *mean* and *standard deviation* given the test results as being:

Stress (N/mm^2)	24.7	25.0	25.3	25.6	25.9
Frequency	5	18	50	20	7

This is shown in fig. 8.3.

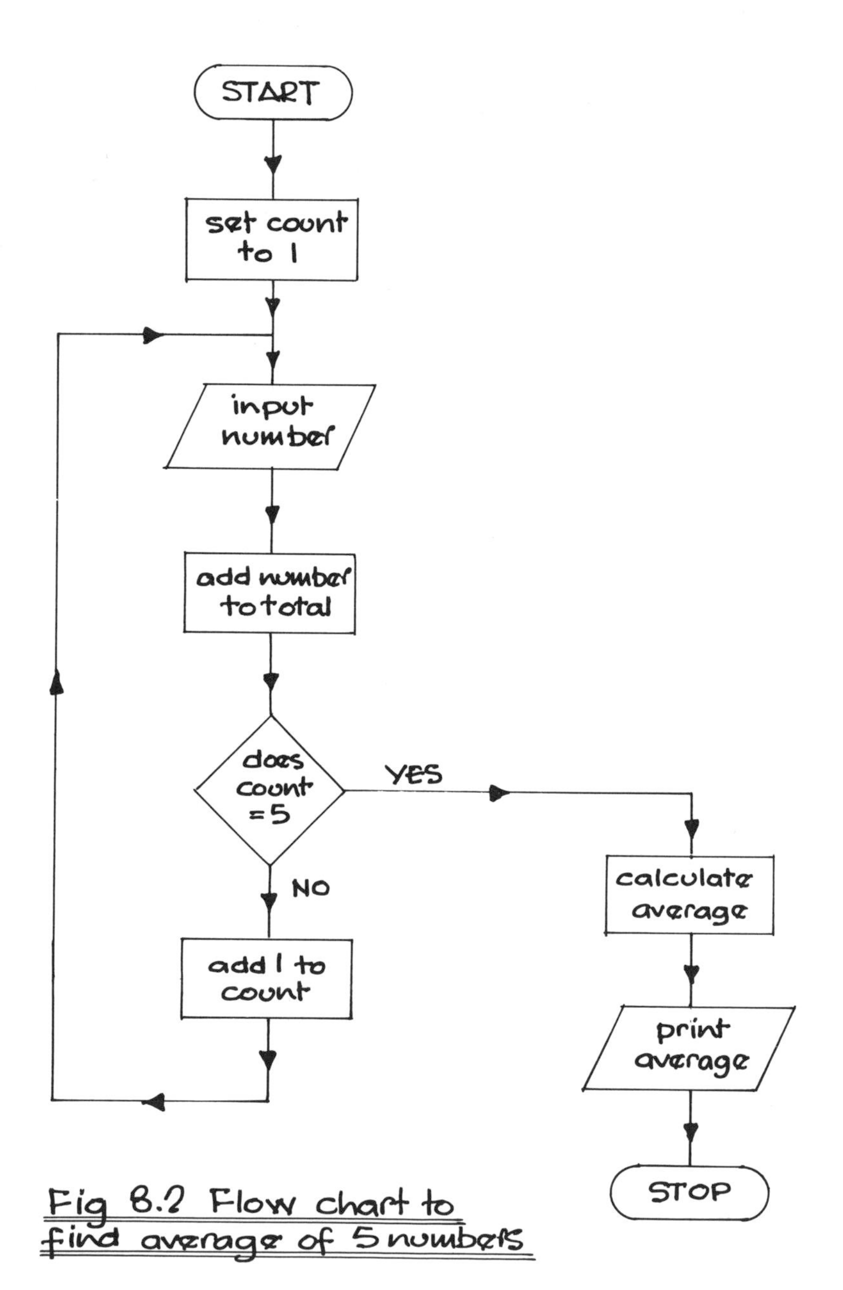

Fig 8.2 Flow chart to find average of 5 numbers

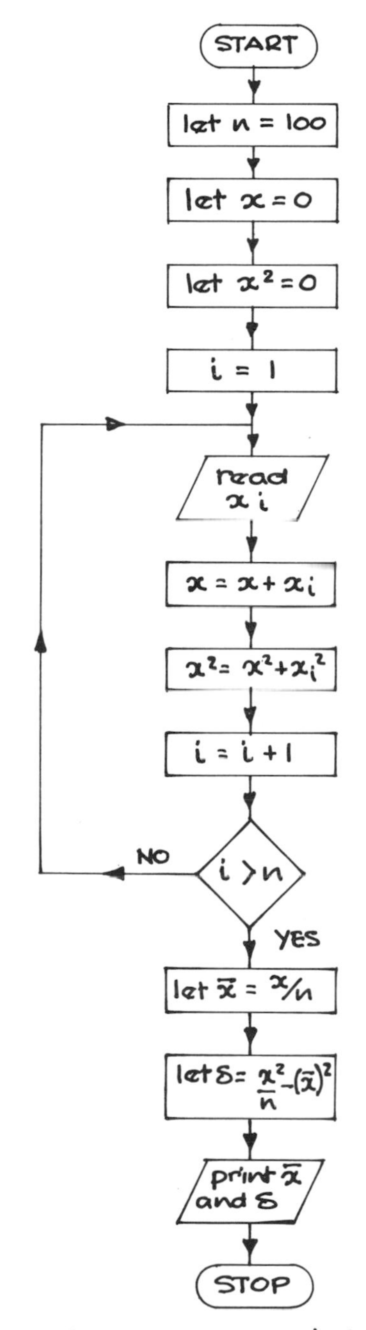

here we are setting n to the number of samples

here we are creating a memory for x and x^2 and setting them at 0

here we are setting up a counting mechanism for the samples

input the first sample

we are now adding all the values

we are now adding together the x^2

having added in that value we now have to move to the next value

we now know how many values so when $i = 100+1$ the computer knows it has all the values

we now create the mean

we now calculate the standard deviation

we now print out the answer

Fig 8.3 Flow chart to solve values for mean and standard deviation for concrete test cubes

PROGRAMMING

Having discussed how the problem is identified, we must now translate the flow chart into a form the computer can understand. There are many *languages* in use in computing today, such as ALGOL, COBOL and FORTRAN, but the one on which we will concentrate is BASIC.

BASIC is an everyday English form of language and can be used with a fairly small number of statements (it was the language used for the example with the programmable calculator) although it can be extended to provide a very powerful language. It is also the language most suitable for *voice command input*, which could take the place of the keyboard in the not-too-distant future.

It would be most convenient if every computer using BASIC had exactly the same commands and instructions, but unfortunately it is necessary to read through the operating manual for each particular system in order to find the details of the language as relevant to that system. Because of this, the commercial software packages are produced for specific systems rather than as an all-purpose pack.

When writing a program, each line has to be given a number. Typically the numbers are increased initially in tens, i.e. 10, 20, 30, 40, ... This allows for amendments to the program to be inserted between lines without having to renumber all the other lines, e.g. 10, 21, 30, 40 . . . . The computer will work through the program in numerical order unless a decision branch tells it to GOTO another line. The program falls into three stages:

(1) *Input*. This is the program and associated data which can be input either each time or from a secondary storage unit (disc or cassette for micro systems).
(2) *Operate*. This is the computer doing the work you have asked it to do.
(3) *Output*. This is the information or solution that you have asked it to produce. As stated earlier, it can be output on to a VDU or to a printer or plotter. Usually the operator will ask first to see the output on the VDU before requesting it to be printed or plotted into *hard copy*.

All programming is done in capitals. The letter O is distinguished from the figure 0 (zero) by using Ø for zero. Most statements have line numbers, but two of them, RUN and LIST, *do not* since they are only commands, i.e. 'RUN' tells the computer to run the program, and 'LIST' lets you list all the stages in the program so that it can be checked and/or amended.

Syntax

Each system has its own syntax. This means that you have to put commas or inverted commas between pieces of information. If you fail to get this right, the computer will display a message, e.g. 'syntax error in line 3Ø'. You will then need to consult the operating manual to correct the syntax.

Program title

If you intend to store a program, you must code it in some way so that you can recall it for future use. The logical thing to do is to give it a title. This is commonly shown as:

1Ø	REM	PROFIT AND LOSS CALCULATIONS
↑	↑	↑
line no.	remarks	Title of program

Symbols for use in BASIC

The following are the symbols most commonly employed, showing the arithmetical and the computer versions:

Arithmetical symbol	Operation	Computer symbol
+	Addition	+
–	Subtraction	–
x	Multiplication	*
÷	Division	/
x^n	Raising to a power	↑ e.g. $x^2 = x↑2$
=	Equal to	=
>	Greater than	>
<	Less than	<
⩾	Greater than or equal to	> =
⩽	Less than or equal to	<=
≠	Not equal to	< >

INPUT AND OUTPUT STATEMENTS

Each BASIC instruction consists of a command to the computer to carry out a certain action. As explained under syntax, a combination of statements can be separated by commas, e.g.

```
20 INPUT A, B, C
```

will tell the computer that you want it to input three numbers. In response it will then display a '?' and expect you to put in the values for A, B and C. To do this, each time the computer displays a '?' you must enter the number and press RETURN. (On most computers, information is typed in and the return key is then pressed to enter it into the computer. Some home computers have dedicated keys, and it is necessary to use these. It is essential, therefore, that you read the operator's manual.) The computer will keep asking for information until it has a value for each of A, B and C. You may wish to check that you have entered the correct values, so if A = 1.9, B = 3.4 and C = 0.4 you request:

```
2Ø READ A, B, C
```

Because this is asking the computer to read certain data, it produces a DATA statement, e.g.

```
3Ø DATA 1.9, 3.4, Ø.4,
```

The computer cannot read fractions so any values involving these must be converted to decimals prior to entry, although you can generate a subroutine to do this. This will require the numerator and denominator to be input.

To output the information, you need to tell the machine to PRINT. A very simple program, the addition of three numbers, is shown below:

```
1Ø  REM  ADDITION OF THREE NUMBERS
2Ø  INPUT A
3Ø  INPUT B
4Ø  INPUT C
5Ø  LET D = A + B + C
6Ø  PRINT D
7Ø  END
```

This may be refined to include, instead

```
60  PRINT  "THE SUM OF A, B, C IS ="
65  PRINT D
```

A further refinement is to use a semicolon to keep the statement to one line, i.e.

```
60  PRINT  "THE SUM OF A, B, C IS ="; D
```

In the first case, if the three values are 5, 7 and 4, the printout would read

```
THE SUM OF A, B, C IS =
16
```

whereas the second refinement would read

```
THE SUM OF A, B, C IS = 16
```

The LET statement

This allows you to introduce the computer to a variable or a specific value, e.g.

```
10  LET X = 30
20  LET D = A + B + C
```

or it can be used for DATA input.

CONDITIONAL AND UNCONDITIONAL BRANCHING

In most problems you may need to repeat a set of instructions. For example, in our case of the cube test results, we need to repeat the instructions until all the test values have been added in. Also all programs need to have a conditional branch if they are going to end and not be continuous loops. In the cube test example we input the 100 test results and once we have reached that stage we want the mean and standard deviation to be output. To do this we use the GOTO and IF . . . THEN statements. Let us look at loops in a simple example.

Example 8.3
In finding the average of five numbers we could write the program:

```
 1Ø  REM  AVERAGE OF FIVE NUMBERS
 2Ø  LET A = 1
 3Ø  LET Z = Ø
 4Ø  INPUT B
 5Ø  LET Z = Z + B
 6Ø  LET A = A + 1
 7Ø  A = 5 THEN 9Ø
 8Ø  GOTO 4Ø
 9Ø  LET Z = Z/A
1ØØ  PRINT  "THE AVERAGE OF THE FIVE NUMBERS
     IS = "; Z
```

In this example the number of times a loop was executed was set by a loop counter. The use of the FOR . . . NEXT statements can make this easier. We can modify the above program to read:

```
 1Ø  REM  AVERAGE OF FIVE NUMBERS
 2Ø  INPUT N
 3Ø  LET Z = Ø
 35  LET A = Ø
 4Ø  FOR A = 1 TO N
 5Ø  INPUT B
 6Ø  LET Z = Z + B
 7Ø  NEXT A
 8Ø  LET Z = Z/N
 9Ø  PRINT  "THE AVERAGE OF ; N NUMBERS IS" = ; Z
1ØØ  END
```

As you can see, the FOR . . . NEXT statement sets up a counter and allows you to set N each time.

STRING VARIABLES

For some problems it is necessary to input store and output variable information which is a mixture of numbers, letters and signs. In order to do this, what are known as *string variables* are stored, for which the $ sign is used. For example, to input the date we say

```
 20  INPUT T $
```

and to print the date later on the output we say

```
250  PRINT T $
```

This can be very useful when producing documents for tendering. There are many more functions, but these are the basic ones and will allow you to develop programs to solve most of the problems in chapters 6 and 7. The main step to take next is to look at setting-out functions such as TAB. This allows you to tabulate your information in a neat, easily readable manner.

Example 8.4
Let us now take our flow chart for the cube tests, shown earlier in fig. 8.3, and convert it to a program. Having solved the particular problem, we have to amend our program to cover any number of tests. The program reads:

```
 10  REM  CUBE TEST ANALYSIS
 20  PRINT  "HOW MANY IN THE TEST"
 30  INPUT N
 40  LET X = 0
 50  LET Y = 0
 60  FOR I = 1 TO N
 70  PRINT  "WHAT WAS THE TEST RESULT"
 80  INPUT A
 90  LET X = X + A
100  LET Y = Y + A↑A
110  NEXT I
120  LET C = X/N
130  LET D = (Y/N - C↑C)*0.5                (D½ = √D)
140  PRINT  "THE MEAN IS ="; C
150  PRINT  "THE STANDARD DEVIATION IS ="; D
160  END
```

If desired, the program can be developed further to calculate how many of the tests fall within one, two and three standard deviations and ultimately to plot the curve and a normal distribution curve.

Exercises

Produce a flow chart and write the program for each of the following:

8.1 To calculate the area of a triangle.

8.2 To calculate the volume of a cone.

8.3 To determine the depreciation of a piece of equipment from the formula

$$D = \frac{P}{t}\left[1 - \left(1 - \frac{r}{100}\right)^t\right]$$

where P = initial cost in £, t = time in years, and r = rate of percentage depreciation per year.

8.4 To calculate the area of an irregular shape using Simpson's rule.

8.5 To produce a Poisson distribution.

Index